Ashoka Reddy Komalla

Innovative Methods for Performance Enhancement of Pulse Oximeters

Ashoka Reddy Komalla

Innovative Methods for Performance Enhancement of Pulse Oximeters

LAP LAMBERT Academic Publishing

Impressum / Imprint
Bibliografische Information der Deutschen Nationalbibliothek: Die Deutsche Nationalbibliothek verzeichnet diese Publikation in der Deutschen Nationalbibliografie; detaillierte bibliografische Daten sind im Internet über http://dnb.d-nb.de abrufbar.
Alle in diesem Buch genannten Marken und Produktnamen unterliegen warenzeichen-, marken- oder patentrechtlichem Schutz bzw. sind Warenzeichen oder eingetragene Warenzeichen der jeweiligen Inhaber. Die Wiedergabe von Marken, Produktnamen, Gebrauchsnamen, Handelsnamen, Warenbezeichnungen u.s.w. in diesem Werk berechtigt auch ohne besondere Kennzeichnung nicht zu der Annahme, dass solche Namen im Sinne der Warenzeichen- und Markenschutzgesetzgebung als frei zu betrachten wären und daher von jedermann benutzt werden dürften.

Bibliographic information published by the Deutsche Nationalbibliothek: The Deutsche Nationalbibliothek lists this publication in the Deutsche Nationalbibliografie; detailed bibliographic data are available in the Internet at http://dnb.d-nb.de.
Any brand names and product names mentioned in this book are subject to trademark, brand or patent protection and are trademarks or registered trademarks of their respective holders. The use of brand names, product names, common names, trade names, product descriptions etc. even without a particular marking in this work is in no way to be construed to mean that such names may be regarded as unrestricted in respect of trademark and brand protection legislation and could thus be used by anyone.

Coverbild / Cover image: www.ingimage.com

Verlag / Publisher:
LAP LAMBERT Academic Publishing
ist ein Imprint der / is a trademark of
OmniScriptum GmbH & Co. KG
Heinrich-Böcking-Str. 6-8, 66121 Saarbrücken, Deutschland / Germany
Email: info@lap-publishing.com

Herstellung: siehe letzte Seite /
Printed at: see last page
ISBN: 978-3-659-46598-7

Zugl. / Approved by: Chennai, IIT Madras, 2008

Copyright © 2015 OmniScriptum GmbH & Co. KG
Alle Rechte vorbehalten. / All rights reserved. Saarbrücken 2015

Innovative Methods for Performance Enhancement of Pulse Oximeters

K. Ashoka Reddy, *Ph.D*

To my family, teachers and well wishers….

Acknowledgement

This book on *Innovative Methods for Performance Enhancement of Pulse Oximeters*, is an outcome based on my research work. I am particularly indebted to my research advisor, Prof. V. Jagadeesh Kumar, *Department of Electrical Engineering, IIT Madras*, India, for introducing me to the field of Pulse Oximetry, providing direction, enthusiasm and encouragement.

I am very grateful to the Management, Kakatiya Institute of Technology & Science, Warangal (KITSW), especially Capt. V. Lakshmikantha Rao, Sri P. Ganga Reddy and Sri P. Narayana Reddy, for giving me an opportunity to pursue research at IIT Madras, India's premier institute. It is my profound duty to express thanks to the All India Council for Technical Education (AICTE), New Delhi and Center for Continuing Education (CCE), IIT Madras for providing financial assistance.

My heartfelt thanks are due to all the volunteers, from whose PPG data the methods proposed in this thesis are validated.

I am glad to have the years of love and affection from a large family consisting of my parents Sri K. Thirupathi Reddy & Smt. K. Anasuya, grandparents, brothers and in-laws, to all of whom I would like to convey my deepest loving gratitude. My wife, Pavani, has been my encouragement, strength and did more than her share of parenting. I am indebted to her for her devotion in bringing up our two enthusiastic sons, Ramakrishna and Chaithanyakrishna.

I express my wholehearted thanks to all my colleagues and friends for their unconditional support.

Preface

Oxygen saturation of arterial blood (SpO_2) is a measure of how much oxygen the arterial blood is carrying as a percentage of the maximum it could carry. Pulse oximeters estimate oxygen saturation utilizing a suitable sensor and a calibration curve obtained through empirical curve fitting. Thus the accuracy of the present day pulse oximeters is affected by sensor and patient dependent parameters.

The book presents basics of pulse oximetry and novel methods for estimation of SpO_2 that are free of interference due to sensor and patient dependent parameters. The book should be of interest to biomedical engineers, healthcare providers. This also could serve to bridge the gap between industrial medical instrumentation application to pulse oximetry and academic research.

This text is arranged into six chapters. Chapter 1 provides introduction to the physiology of oxygen transport in blood, oxygen saturation of arterial blood, photoplethysmography, principles of pulse oximetry, motivation and objectives.

Chapter 2 presents a novel model based method for the computation of SpO_2 utilizing the traditional negative feedback compensation scheme that normalizes the red and IR PPG signals. A prototype pulse oximeter that employs a sensor housing red and infrared (IR) LEDs and suitable photo detectors is developed to validate the proposed method. The developed prototype employs the audio channel of a PC for data acquisition dispensing with expensive analog to digital converter hardware. The proposed pulse oximeter estimates SpO_2 using an analytical expression and hence does not rely on calibration curves.

In Chapter 3, a novel method of estimation of SpO_2 wherein the red and IR PPG signals are processed appropriately with a view to remove the source and detector dependent parameters is presented. Appropriate expressions are derived based on a model of light propagation through tissue, bone and blood such that the equation for the computation of SpO_2, utilizing the processed PPG signals and the well-known extinction coefficients of HbO_2 and Hb, is devoid of all interfering parameters, thus obviating the need for extensive calibration.

Chapter 4 deals with a refined method of measurement of SpO_2 employing only the peak-to-peak values of the suitably processed red and IR PPG signals. This method too does not require extensive calibration needed in present day pulse oximeters that employ empirical curve fitting.

Chapter 5 starts with a brief review of existing motion artifact reduction methods applied to PPG signals and presents a method to reduce the motion artifacts from the corrupted PPG signals utilizing singular value decomposition (SVD). Application of SVD technique leads to a stable and reliable estimation of SpO_2 even when the PPG signals are distorted by motion artifacts.

Chapter 6 proposes a simple but effective method for reduction of motion artifacts using the well known Fourier series analysis (FSA) applied to a PPG signal on a cycle-by-cycle basis. Over and above artifact reduction, the FSA also provides data compression. Practical applicability of all the techniques outlined in this book is verified by experiments conducted using suitable prototype units.

I wish to thank Lambert Academic Publishing for moving this manuscript toward publication.

I would welcome suggestions for subsequent editions.

K. Ashoka Reddy

Department of Electronics & Instrumentation Engineering

Kakatiya Institute of Technology &Science, Warangal

Telangana – 506015, India

January 2015

kar@eie.kitsw.ac.in

Table of Contents

Acknowledgement	i
Preface	ii
Table of Contents	iv
List of Tables	vi
List of Figures	vii
Abbreviations	x

1 Introduction ... 1

    1.1 Physiological Signals for Diagnostics ... 1
    1.2 Blood and its Composition ... 1
    1.3 Systemic and Pulmonary Circulation ... 2
    1.4 Mechanism of Oxygen Exchange ... 2
    1.5 Functional and Dysfunctional Hemoglobin ... 4
    1.6 Oxygen Saturation ... 5
        1.6.1 Measurement of Oxygen Saturation (Oximetry), 7
        1.6.2 Arterial Blood Gas Analyzer and CO-Oximeter, 8
        1.6.3 Photoplethysmography, 9
        1.6.4 Pulse Oximetry, 11
        1.6.5 Principle of Operation of a Pulse Oximeter, 14
    1.7 Problems Associated with Pulse Oximetry ... 18
    1.8 Limitations of Pulse Oximetry in Vogue and Motivation for Research ... 21
    1.9 Objectives and Scope ... 23
    1.10 Organization of the Book ... 23

2 Novel Feedback Compensation Based Method for the Measurement of SpO_2 ... 25

    2.1 Existing Methods of Pulse Oximetry - Problems and Challenges ... 25
    2.2 Feedback Compensation for Normalization of a PPG ... 26
    2.3 Novel Method of Computation of SpO_2 Utilizing the Normalized PPG Signals ... 31
    2.4 Possible Sources of Errors ... 33
    2.5 Experimental Setup and Results ... 36

3 A Novel Slope Based Method of Measurement of SpO_2 ... 44

    3.1 Modeling Light Propagation through Test Object (Tissue) ... 44
    3.2 The Slope Based Method of Measurement of SpO_2 ... 48
    3.3 Experimental Results ... 51
        3.3.1 Experimentation to Illustrate the Efficacy of the Proposed Method in Removing the Influence of Source Intensity ... 53

		3.3.2 Experimentation to Illustrate the Efficacy of the Proposed Method in Removing the Influence of Detector sensitivity	55
4	**A Novel Peak Value Based Method of Measurement of SpO_2**		**66**
	4.1	Introduction	66
	4.2	The Peak Value Method of Estimation of SpO_2	66
		4.2.1 Error in the Proposed Method of Measurement of SpO_2, 68	
	4.3	Experimental Results	68
5	**Motion Artifact Reduction in PPG Signals using Singular Value Decomposition**		**72**
	5.1	SVD for Motion Artifact Reduction	75
		5.1.1 Singular Value Decomposition, 75	
		5.1.2 Principal Component Extraction using SVD, 76	
	5.3	Experimental Results	77
6	Fourier Series Analysis for Motion Artifact Reduction and Data Compression of PPG Signals		88
	6.1	Need for Alternative Method for Motion Artifact Reduction	88
	6.2	Cycle-by-cycle FSA Applied to PPG Signals	88
	6.3	Cycle-by-cycle FSA for Motion Artifact Reduction	93
	6.4	Experimental Results	96
7	**Summary and Conclusions**		**105**
	7.1	Summary of the Work Presented in this Book	105
	7.2	Conclusions	106
	7.3	Scope for Future Work	108

References	**109**
Annexure 1	**118**
Annexure 2	**121**
Annexure 3	**124**
Annexure 4	**125**

List of Tables

Table No.	Caption	Page
Table 2.1	Error in the computation of SpO_2 due to mismatch in the DC voltages of red and IR PPG signals	34
Table 2.2	Error in the computation of SpO_2 due to errors in choosing extinction coefficients	35
Table 2.3	Extinction coefficients employed in the prototype	39
Table 2.4	Comparison of SpO_2 values obtained with the prototype and a CPO	42
Table 2.5	SpO_2 computed using proposed method on L&T data	43
Table 3.1	Effect of processing on the source intensity of the PPG signals obtained from a volunteer.	57
Table 3.2	Effect of detector sensitivity processed PPG signals	59
Table 3.3	SpO_2 computed using slope based method	65
Table 4.1	SpO_2 computed using peak value based method	69
Table 4.1	SpO_2 computed using peak value method on L&T data	71
Table 6.1	Fourier coefficients of sample PPG cycles of Fig. 6.1(a)	92
Table 6.2	Error in reconstruction of PPG as a function of chosen number of Fourier coefficients for reconstruction	94
Table 6.3	Effectiveness of FSA method in restoring peak-to-peak values of a PPG	102

List of Figures

Figure No.	Caption	Page
Fig. 1.1	Sensor for obtaining a PPG signal utilizing the transmitted light through finger	10
Fig. 1.2	Components of a typical PPG signal	10
Fig. 1.3	Absorption spectra of Hb and HbO_2	16
Fig. 1.4	A typical pulse oximeter calibration curve showing empirical relationship between actual SaO_2 and normalized ratio R	17
Fig. 2.1(a)	Reflectance type PPG sensor head	27
Fig. 2.1(b)	Transmission type PPG sensor head	27
Fig. 2.2	Feedback compensation scheme for obtaining a quantitative PPG	29
Fig. 2.3	Absorption spectra of Hb and HbO_2	32
Fig. 2.4	Analog front-end for obtaining red and IR PPG signal utilizing feedback compensation	37
Fig. 2.5(a)	PCB of the analog signal processing part for processing PPG signals	40
Fig. 2.5 b)	The developed photoplethysmograph for pulse oximetry	40
Fig. 2.6	Snapshot of the front panel of the developed instrument	41
Fig. 3.1	A model showing possible light interactions with cells with trans-illuminated object (finger)	45
Fig. 3.2(a)	Optical attenuation through a cell (tissue)	46
Fig. 3.2(b)	Optical attenuation through a cell-equivalent representation	46
Fig. 3.3	Process of obtaining slope information from a PPG signal	50
Fig. 3.4	Analog front-end for obtaining PPG signals with different source intensities and detector sensitivities	52
Fig. 3.5	Complete experimental setup showing developed prototype PPG unit and data acquisition facility	54
Fig. 3.6	Snapshot of the front panel of the prototype PPG, data recorded from a volunteer	56
Fig. 3.7	PPG signals recorded simultaneously at different source intensities and the overlapped PPG signals after applying natural logarithm	58
Fig. 3.8	PPG signals recorded at different detector gain settings and overlapped PPG signals after applying natural logarithm	60

Fig. 3.9	Eighteen cycles of a processed PPG data with their linear portions identified	61
Fig. 3.10	Linear regression performed on the samples of identified linear portions of PPG cycles to obtain slope information	62
Fig. 3.11	Snapshot of the front panel of the prototype pulse oximeter employing the proposed slope based method	64
Fig. 4.1	Snapshot of the front panel of the prototype pulse oximeter based on the proposed peak value method	70
Fig. 4.2	Estimated SpO_2 obtained from traditional method, readings of CPO and proposed peak value based method	70
Fig. 5.1	(a) Artifact-free red PPG as acquired (b) Recovered red PPG using the proposed SVD technique (c) SpO_2 computed from the original and recovered PPG signals	79
Fig. 5.2	SVR spectrum obtained during the reconstruction of first cycle of PPG signal shown in Fig. 5.1(a)	79
Fig. 5.3	SVR spectra of the six PPG cycles shown in Fig. 5.1(a)	81
Fig. 5.4	(a) Artifact corrupted PPG due to horizontal motion of finger (b) Recovered PPG using proposed SVD method	83
Fig. 5.5	(a) Artifact induced PPG by pressurizing the probe (b) Recovered PPG using proposed SVD based method	83
Fig. 5.6	(a) Artifact corrupted PPG due to vertical motion of finger (b) Reconstructed PPG using proposed SVD based method	84
Fig. 5.7	(a) Artifact corrupted PPG due to bending of finger (b) Reconstructed PPG using proposed SVD based method	85
Fig. 5.8	(a) Artifact corrupted IR PPG as obtained (b) Recovered PPG using the proposed technique (c) SpO_2 computed from the original and recovered PPG signals	86
Fig. 5.9	SVR profiles of the seven cycles of artifact corrupted IR PPG signal shown in Fig. 5.8(a)	87
Fig. 6.1	(a) Sample PPG chosen with a high heart rate variability (HRV) for testing the efficacy of the proposed method (b) Reconstructed PPG obtained after applying FSA on (a)	90
Fig. 6.2	*NMRSE* (%) as a function of number of Fourier coefficients	94
Fig. 6.3	(a) PPG corrupted with motion artifact, chosen to test proposed FSA mehtod's motion artifact reduction capability (b) Clean PPG recovered from corrupted PPG of (a)	96
Fig. 6.4	(a) PPG signal corrupted by horizontal motion of finger (b) Output of MA filter (c) Output utilizing the FSA method	98
Fig. 6.5	(a) PPG signal contaminated by pressurizing the probe (b) Processed PPG signal obtained from the MA method (c) Result after processing using the FSA method	98
Fig. 6.6	(a) PPG signal corrupted with artifact due to vertical motion	

	of finger (b) Signal obtained after applying the traditional MA method (c) Signal obtained after applying FSA method	100
Fig. 6.7	(a) Artifact induced in PPG signal by bending of the finger (b) Processed signal using MA filter (c) Processed signal using the proposed FSA method	101
Fig. 6.8	(a) and (c) show red and IR PPG signals respectively with artifact induced by waving motion of the finger (b) and (d) show the red and IR PPG signals respectively recovered using FSA method (e) SpO_2 computed using the signals before and after applying the FSA method	103

Abbreviations

BPM	Beats per minute
COHb	Carboxy-hemoglobin
CPO	Commercial pulse oximeter
$\langle Hb \rangle$	Concentration of Hb
$\langle HbO_2 \rangle$	Concentration of HbO$_2$
Hb	Deoxy-hemoglobin
ε_{HbR}	Extinction coefficient of deoxy-hemoglobin at red wavelength
ε_{HbIR}	Extinction coefficient of deoxy-hemoglobin at IR wavelength
ε_{HbOR}	Extinction coefficient of oxy-hemoglobin at red wavelength
ε_{HbOIR}	Extinction coefficient of oxy-hemoglobin at IR wavelength
FSA	Fourier series analysis
HR	Heart rate
HRV	Heart rate variability
ICA	Independent component analysis
IR	Infrared
LED	Light emitting diode
MA	Moving average
MetHb	Met-hemoglobin
ln	Natural logarithm (base e)
$NRMSE$	Normalized root mean square error
HbO$_2$	Oxy-hemoglobin
SaO$_2$	Oxygen saturation of arterial blood
SpO$_2$	Oxygen saturation as indicated by pulse oximeter
SvO$_2$	Oxygen saturation of venous blood

P_{CO_2}	Partial pressure of carbondioxide
P_{O_2}	Partial pressure of oxygen
V_{pR}	Peak-to-peak amplitude of a processed PPG cycle at red wavelength
V_{pIR}	Peak-to-peak amplitude of a processed PPG cycle at IR wavelength
PPG	Photoplethysmogram
PR	Pulse rate
$PPG_{R\lambda}$	Reflectance type PPG obtained at a wavelength λ
$\sigma_1,...,\sigma_n$	Singular values
SVD	Singular value decomposition
SVR	Singular value ratio
m_R	Slope of the linear portion of a processed red PPG cycle
m_{IR}	Slope of the linear portion of a processed IR PPG cycle
SD	Standard deviation
$PPG_{T\lambda}$	Transmission type PPG obtained at a wavelength λ
VI	Virtual instrument
λ	Wavelength

1. Introduction

1.1 Physiological Signals for Diagnostics

Health and physical condition of the human body can be ascertained by making suitable measurements on the electrical, chemical and acoustic signals emanating from it [1]-[3]. Any malfunction due to a disease or injury causes one or more of these signals to differ from their expected normal form. Quite often information reflecting the functioning or malfunctioning of the underlying biological system is entwined in a complex manner in these signals and hence such information has to be extracted from these signals.

The four traditional vital signs, normally used by medical doctors all over the globe, to obtain critical knowledge about a patient's state of health are the pulse rate, the respiratory rate, the body temperature and the blood pressure. For life, nothing is more important than oxygen supply to the body. Oxygen required for all parts of the body is carried by the arterial blood.

1.2 Blood and its Composition

The arterial blood carries nutrients and oxygen to all the cells of the body and removes waste from the cells to waste disposal parts of the body. About 55 % of blood is composed of a liquid called the plasma, 43% of the blood is made of red blood cells (RBCs, also known as erythrocytes), 1.5% white blood cells (leukocytes), and 0.5% platelets (thrombocytes). 90% of plasma is made of water with some proteins and other chemicals dissolved in it. On the other hand RBCs are mainly made of hemoglobin molecules. Hemoglobin is responsible for the transport of oxygen to various other cells of the body [4]. Typically each mm^3

of blood contains nearly six million RBCs and each RBC is made of about 280 million hemoglobin molecules. Normally hemoglobin concentration in whole blood is between 134 and 173 g/l [5]. The natural chromophore contained in the hemoglobin absorbs light depending on the number of oxygen molecules attached to a molecule of hemoglobin. If the hemoglobin molecule is saturated with oxygen then the natural chromophore absorbs all other components of light except the red and that gives the arterial blood its red color. Blood is circulated through out the body by the systemic and pulmonary circulation system [6].

1.3 Systemic and Pulmonary Circulation

The flow of arterial blood, the replenishment of oxygen and exhaling of carbondioxide are controlled by the cardio-pulmonary system, comprising the heart, lungs and the blood vessels (arteries, veins and capillaries). Arteries (except the pulmonary artery, which carries oxygen depleted blood to the lungs for replenishment of oxygen) carry oxygenated blood from the heart to all parts of the body. Arteries terminate into capillaries and the blood in the capillaries provides oxygen and nutrients to the cells and picks up the waste including carbondioxide from the cells. The capillaries terminate to small veins and the small veins lead to bigger veins. Thus oxygen depleted blood containing additional waste is brought back to the right atrium of the heart through these veins. This blood then passes to right ventricle and gets pushed through the pulmonary artery to the lungs. In the lung capillaries, the exhale of carbondioxide and infusing of oxygen takes place and the oxygen-rich blood from the lungs returns back through the pulmonary veins to the heart, thus completing one cycle.

1.4 Mechanism of Oxygen Exchange

Air inhaled during breathing enters the lungs. Oxygen in the air is trapped by the hemoglobin molecules and carbondioxide from the blood is released to the air that gets exhaled [7]. A heterogeneous collection of gas exchange units called alveoli in the lungs surrounded by large pulmonary capillary beds aid this gas exchange through diffusion. Diffusion of a gas requires differential partial pressure. The inhaled air (at near sea level) is at a total pressure of 101 kPa and is made up of 21% of oxygen, 78% nitrogen (N_2) and small quantities of carbondioxide, argon and helium. The partial pressures exerted by the two main gases added together nearly equal the atmospheric pressure. The partial pressure of oxygen (P_{O_2}) of dry air at sea level is therefore approximately 21 kPa but by the time air passes through the trachea and reaches the alveoli, the P_{O_2} falls to about 13 kPa. Blood returning to the lungs has a P_{O_2} of 5 kPa. A thin wall (about 0.5 μm thick) between the pulmonary capillaries and the alveoli permits diffusion of gases. Since P_{O_2} of air in the alveoli is 13 kPa and in the pulmonary capillaries is 5 kPa, oxygen diffuses from alveoli to the blood in the pulmonary capillaries. On the other hand, the partial pressure gradient for carbondioxide ensures the diffusion of carbondioxide from the blood to the air trapped in the alveoli. Normally partial pressure of nitrogen in the blood and the alveolar air is about the same hence very little nitrogen is diffused in either direction.

Thus the blood returning to the left side of the heart to be pumped into the systemic circulation is replenished with oxygen. If this process is normal then the P_{O_2} of pulmonary venous blood would be equal to the P_{O_2} in the alveoli. Any malfunction would render the pulmonary vein P_{O_2} to be less than the P_{O_2} in alveoli, resulting in reduced amount of oxygen in the arterial blood. In fact, the amount of oxygen bound to the hemoglobin at any time is dictated by the P_{O_2} to which the hemoglobin is exposed. When the arterial blood enters body tissue through capillaries, wherein the P_{O_2} is lower than the arterial P_{O_2}, oxygen is detached from the hemoglobin and enters the tissue. The total quantity of oxygen bound to hemoglobin in normal arterial blood is approximately 19 ml per 100 ml of blood at a P_{O_2} of 13 kPa. On passing through tissue capillaries this amount is reduced to 14 ml per 100 ml of blood at a P_{O_2} of 5 kPa. Thus under normal conditions, about 5 ml of oxygen is consumed by tissues from each 100 ml of blood that passes through tissue capillaries during each cycle. When the blood returns to the lungs, approximately 5 ml of oxygen diffuses from alveoli into each 100 ml of blood.

Nearly 98.5% of the diffused oxygen gets bounded to hemoglobin molecules and the remaining 1.5% gets dissolved in plasma. Each hemoglobin molecule is made up of four "heme" (the iron containing portion of hemoglobin) groups and a protein group, known as "globin" (amino acid chains that form a protein). Each heme unit can carry one oxygen molecule and hence one molecule of hemoglobin can carry up to four molecules of oxygen. When one hemoglobin molecule bounds four oxygen molecules, it becomes fully saturated with oxygen. Each gram of fully saturated hemoglobin then contains 1.3 ml of oxygen. However, not all hemoglobin molecules participate in oxygen transport.

1.5 Functional and Dysfunctional Hemoglobin

Blood contains several forms of hemoglobin, of which some are useful in oxygen transport and some are not. Functional hemoglobin are those that are capable of carrying oxygen and include hemoglobin bounded with oxygen molecules, called oxyhemoglobin (oxygenated hemoglobin, HbO_2) and hemoglobin not bounded with any other molecule called reduced hemoglobin (deoxy-hemoglobin, Hb). Hemoglobin which are incapable of carrying oxygen are called *dysfunctional hemoglobin* (dyshemoglobin). These are hemoglobin bounded with molecule(s) other than oxygen. They include carboxyhemoglobin (COHb) and methemoglobin (MetHb). COHb is formed when carbonmonoxide (CO) bonds to hemoglobin. COHb exits in varying degrees as a consequence of smoking and urban pollution. The level of COHb may become as high as 45% as a result of smoke inhalation. MetHb is oxidized hemoglobin, normally less than 1 % of the total hemoglobin. MetHb is not capable of binding oxygen and hence cannot aid in oxygen transport. Under normal conditions, HbO_2 and Hb amount to 99 % of the total hemoglobin present in the blood.

1.6 Oxygen Saturation

Whether a person is sleeping, resting or active, every part in the person's body requires oxygen. The amount of oxygen required for a particular part of the body depends on the degree of activity of that part but is never zero. While some part of the body can tolerate deprivation of oxygen for limited periods of time, many vital organs will become irreversibly damaged when not supplied with proper amount of oxygen, even for a very short period. Of the body organs, the brain is by far the most sensitive to oxygen deficit. Then it is absolutely necessary that the amount of oxygen carried by the arterial blood is measured to estimate the level of functioning of various parts of the cardio-pulmonary system.

The direct method of measurement of oxygen content in arterial blood is to do a complete analysis to ascertain the various concentrations of gasses in arterial blood. Such an analysis would require drawing of blood directly from an artery and hence requires the services of a competent surgeon. Alternate methods for the determination of the gas contents of arterial blood without the need for puncturing and drawing blood from an artery have been proposed [8]. One such method is *pulse oximetry*, wherein the amount of oxygen in arterial blood is indirectly measured in terms of *oxygen saturation in arterial blood* as a percentage [9], [10].

The oxygen saturation in arterial blood is a measure of how much oxygen the arterial blood is carrying as a percentage of the maximum it could carry. Hemoglobin is the oxygen carrying agent of blood and one molecule can carry four oxygen molecules. If in total there are N hemoglobin, then

N Hemoglobin + N Oxygen = 25% saturated hemoglobin

N Hemoglobin + 2 N Oxygen = 50% saturated hemoglobin

N Hemoglobin + 3 N Oxygen = 75% saturated hemoglobin

N Hemoglobin + 4 N Oxygen = 100% saturated hemoglobin

Thus if each and every hemoglobin of arterial blood carries four oxygen molecules, the arterial blood is fully (100 %) saturated with oxygen. Or in other words, if N_{Hb} is the number of hemoglobin molecules and N_{O_2} is the number of oxygen bounded with hemoglobin in arterial blood then we can define oxygen saturations as

$$\text{Oxygen saturation (SaO}_2\text{)} = \frac{N_{O_2}}{4N_{Hb}} \times 100 \% \qquad (1.1)$$

Oxygen saturation is normally expressed as a percentage rather than as a ratio. If we represent the concentration of oxyhemoglobin as $\langle HbO_2 \rangle$ and the concentration of deoxyhemoglobin as $\langle Hb \rangle$ in arterial blood then equation (1.1) can be rewritten as [11]

$$SaO_2 = \frac{\langle HbO_2 \rangle}{\langle Hb \rangle + \langle HbO_2 \rangle} \times 100\% \qquad (1.2)$$

Rearranging equation (1.2) results in

$$SaO_2 = \frac{\frac{\langle HbO_2 \rangle}{\langle Hb \rangle}}{1 + \frac{\langle HbO_2 \rangle}{\langle Hb \rangle}} \; 100\% = \frac{Q}{1+Q} \; 100\% \; , \qquad (1.3)$$

where $Q \left(= \frac{\langle HbO_2 \rangle}{\langle Hb \rangle} \right)$ is the ratio of HbO_2 to Hb in arterial blood. Equation (1.2) assumes that dysfunctional hemoglobin are negligible. If both functional and dysfunctional hemoglobin are present in the arterial blood then oxygen saturation is expressed as a *fractional oxygen saturation* and is given by

$$\text{Fractional } SaO_2 = \frac{\langle HbO_2 \rangle}{\langle Hb \rangle + \langle HbO_2 \rangle + \langle COHb \rangle + \langle MetHb \rangle} \times 100\% \qquad (1.4)$$

Under normal physiological conditions arterial blood is 97% saturated with oxygen. A healthy, nonsmoking person should have arterial oxygen saturation between 94% and 100%. Anything below 90% could quickly lead to life threatening complications. Saturations lower than 90% may be caused by chronic obstructive pulmonary disease (COPD), excessive bleeding, smoking and malfunctioning blood vessels, especially capillaries. Once oxygen is supplied to various parts of the body, the returning blood in the veins will be depleted of oxygen. The functional oxygen saturation of venous blood (SvO_2) is about 75% [12].

1.6.1 Measurement of Oxygen Saturation (Oximetry)

Three types of oximetry are now in clinical use [13]:

a)   Invasive (*in-vitro*) CO-Oximetry and arterial blood gas analysis.

b)   Invasive fiber optic based oximetry to determine oxygen saturation in arterial flow, mixed arterial-venous flow or intracardiac flow.

c)   Noninvasive pulse oximetry to monitor arterial oxygen saturation at any part of the body.

1.6.2 Arterial Blood Gas Analyzer and CO-Oximeter

Prior to the widespread use of the present day noninvasive pulse oximeter, the arterial blood gas (ABG) analysis and CO-Oximetry were the main methods employed for the measurement of arterial oxygen saturation. As its name implies, the ABG test is conducted by taking a blood sample from an artery and performing complete gas analysis. For this purpose either the radial artery at the wrist or the brachial artery at the elbow would have to be punctured. The common practice was to draw the samples at regular intervals, several times a day or even several times an hour, and analyze using *in-vitro* blood gas analyzer. The AGB analyzer gives a full picture of blood including pH, Po_2 and Pco_2, the bicarbonate concentration in addition to the SaO_2.

CO-Oximeter or haemoximeter calculates the actual concentrations of the Hb, HbO_2, COHb and MetHb but again requires a sample of blood drawn from an artery. It works on the spectrometric principles using four different wavelengths of light and measures the fractional SaO_2. The CO-Oximeter is one of the most accurate methods available for measuring the four clinically relevant hemoglobin species. It is the "gold standard" against which other methods of measurement are compared [14].

However, both these methods are invasive and risky as puncturing an artery may result in spasm, excessive bleeding, vessel obstruction and infection [15].

Because both these methods are time consuming, invasive and provide accurate SaO_2 readings only at the times at which the samples are drawn, *in-vitro* continuous and noninvasive methods are developed. One such noninvasive method is *pulse oximetry*. Pulse oximetry has now become the most popular method employed for the determination of SaO_2. When SaO_2 is measured using the principle of pulse oximetry, it is customary to indicate it as SpO_2. In the method of pulse oximetry SpO_2 is computed using a couple of photoplethysmographs.

1.6.3 Photoplethysmography

Photoplethysmography, a non-invasive electro-optic method developed by Hertzman, provides information on the blood volume flowing at a particular test site on the body close to the skin [16]. A Photoplethysmogram (PPG) is obtained by illuminating a part of the body of interest and acquiring either the reflected or transmitted light. For obtaining PPG by way of detecting the transmitted light, we place a light source of wavelength λ having a constant intensity $I_{IN\lambda}$ on one side of an extremity, say finger tip, and detect the transmitted light through the finger by a suitable photo detector placed on the side opposite to that of the source as indicated in Fig. 1.1. A typical PPG signal, shown in Fig. 1.2, is made of a large DC component that is due to a large part of light from the source passing through skin-muscle-bone without coming into contact with blood vessels at all and reaching the photo detector, a very small component having a very low frequency due to light from the source passing through, apart from the skin-muscle-bone, the venous blood and a much smaller component at the frequency of the heart beat due to light from the source also passing through arterial blood vessels. Just after the systole, blood volume increases in the arteries thereby reducing the received light intensity. During diastole, blood volume in the arteries decreases and hence an increase in the light transmission. Thus the part of detected signal due to the arterial blood appears pulsatile in nature at the heart rate, as shown in

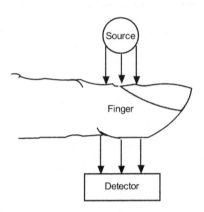

Fig. 1.1 Sensor for obtaining a PPG signal utilizing the transmitted light through finger

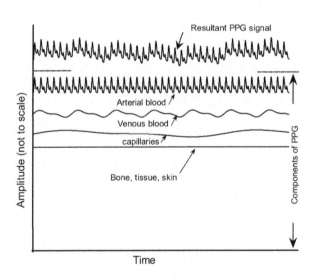

Fig. 1.2 Components of a typical PPG signal

Fig. 1.2. At a classical measuring site, about 99% of the detected signal comes from skin-tissue-bone, 0.9% from venous blood and about 0.1% from arterial blood volume. The pulsatile portion of the PPG arises due to the light passing through arterial blood and hence has information contained in the arterial blood flow; heart rate, heart rate variability, respiration and blood pressure to name a few [17]. Similarly the slow varying component of a PPG is due to the venous blood. It has been shown that utilizing the slow varying component of a single wavelength PPG, diagnosis and monitoring of peripheral vascular hemo-dynamics and venous dysfunction (thrombosis) can be achieved [18], [19]. In the method of pulse oximetry, using two PPG signals, one obtained using a light source in the red wavelength region and the other in the infrared region, the level of oxygen saturation in arterial blood is ascertained non-invasively.

1.6.4 Pulse Oximetry

Measurement of oxygen saturation using photoplethysmographic principle dates back to the 1930s, when Karl Matthes, who is regarded as the Father of oximetry, established that red light can pass through oxyhemoglobin but reduced hemoglobin absorbs it and built the first device to continuously measure blood oxygen saturation *in vivo* by transilluminating tissue. He used two wavelengths of light, one of which was sensitive to oxygen saturation and the other, which was in infrared range, was used to compensate for changes in tissue thickness and light intensity. Although useful in following the trends in saturation, difficulties with calibration made it an unwieldy device. In the early 1940s, Millikan devised an instrument and coined the term "oximeter" to measure arterial oxygen saturation from the ear of a pilot [20] during World War II, so as to regulate the oxygen delivery system to help pilots flying at high altitudes in pressurized cockpits. His ear oximeter was not calibrated and one had to guess the normal saturation for each subject. Most important subsequent works were performed by Goldie [21], Wood and Geraci [22], that resulted in the improvement of Millikan's ear oximeter. In 1949, Brinkman and Zijlstra [23] were first to describe the monitoring of SaO_2 based upon skin reflectance spectroscopy from forehead, first *in-vitro*, and then *in-vivo*. Their innovative idea to use light reflection instead of tissue transillumination resulted in monitoring of SaO_2 from virtually any part of the body. This was followed by a photoelectric method proposed by Sekelj *et al.* [24] for SaO_2 determination. In 1960, Polanyi and Hehir [25] developed the fiber optic catheter oximeter which is the basis for the modern invasive oximeter. In 1964, a surgeon, Robert Shaw, built a self calibrating ear oximeter, using eight wavelengths between 650 nm and 1050 nm, to identify and separate Hb species including COHb and MetHb. In early 1970s Hewlett-Packard improved his method and released the first commercial eight-wavelength ear oximeter (HP 47201A). This oximeter used a sensor mounted on the ear with light delivered via a fiber optic cable and used a heating element to keep the tissue locally perfused with blood. Meanwhile, Cohen and Wardsworth added significant advancements in noninvasive reflectance oximetry [26]. In 1972, Takuo Aoyagi, an engineer working with Nihon Kohden Corporation in Tokyo, Japan, invented the present day two-wavelength pulse oximetry [27]. Aoyagi while trying to develop a noninvasive method to determine cardiac output using *cardiogreen* dye, measured light passing through the earlobe, noted that the light transmitted through the earlobe exhibited pulsatile variations. While attempting to eliminate these variations, he discovered that the ratio of pulsatile signals measured at two different wavelengths could be related to the oxygen levels of the arterial blood. Aoyagi's method also eliminated the requirement to know the intensity of light that was entering the tissue-under-test. Aoyagi and his team announced the first pulse oximeter in March 1974, the OLV-5100. This was a significant development because it reduced the number of wavelengths necessary for measurement of SpO_2 from the eight used in the HP instrument down to two [28]. However, Aoyagi's device used a tungsten light source and two narrow band filters to generate the required two channels of nearly monochromatic light. Unfortunately, these filters also blocked majority of the light intensity from the source, leaving precious little light for measurement of oxygen saturation. However, all these early instruments suffered from one or more of the following drawbacks [29]:

a) Lack of adequate calibration procedures.

b) Difficulty in differentiating tissue, arterial blood and venous blood.

c) Error due to unknown optical path length.

In the late 1970s, several groups began developmental work in pulse oximetry using Aoyagi's idea and fingertip probes were introduced. With subsequent developments in semiconductor technology, the solid state devices such as LEDs, photodiodes and microprocessors steered the current era of modern pulse oximetry. LEDs generated required narrowband light with controlled wavelengths, exactly the type of light required to vastly improve the signal quality of oximeters. In 1981, Nellcor and Ohmeda (now GE) introduced commercial pulse oximeters utilizing small LEDs and photodiode mounted directly on the sensor probe applied to the patients. Today, there are many manufacturers producing pulse oximeters with elevated levels of confidence in the readings of oxygen saturation [30].

In course of time, pulse oximetry has revolutionized the concept of clinical monitoring since electrocardiography. On 1st January 1990, the American Society of Anesthesiologists (ASA) made pulse oximeter a standard for intraoperative monitoring [31]. Since then, pulse oximetry has become the standard technique for monitoring oxygenation during: procedural sedation, anesthesia, post anesthesia care unit, intensive care unit including neonatal intensive care unit, and recovery from anesthesia.

1.6.5 Principle of Operation of a Pulse Oximeter

Pulse oximeters derive their name due to the fact that they operate on the pulsatile portions of red and IR PPG signals to estimate the oxygen saturation in arterial blood. It is seen from equation (1.2), to compute SaO_2, the concentrations of oxy and deoxyhemoglobin ($\langle HbO_2 \rangle$ and $\langle Hb \rangle$) must be known. All the present day pulse oximeters utilize Beer-Lambert's law, which states that concentration of an absorbing substance in a solution can be determined from the intensity of light, at a specific wavelength, transmitted through the solution. Here, the light intensity of transmitted light (I_0) is related to the light intensity of incident light (I_{IN}) by:

$$I_0 = I_{IN} e^{-\varepsilon_\lambda cL} \qquad (1.5)$$

where ε_λ is the wavelength dependent extinction coefficient (normally expressed in $l.mmol^{-1}.cm^{-1}$), c is the concentration of the absorber ($mmol.l^{-1}$) and L is the optical path length (cm). The light absorbed while passing through the solution is expressed in terms of absorbance, given by:

$$A = \ln\left(\frac{I_{IN}}{I_0}\right) = \varepsilon_\lambda cL \qquad (1.6)$$

where A is the absorbance, a dimensionless quantity, normally termed the optical density (OD). Hence Beer-Lambert's law allows us to determine the unknown concentration if the absorbance of the light is measured and extinction coefficient at that wavelength and optical path length are known. If multiple absorbers are present in the path of light then each absorber contributes its part and the resulting total absorbance can be expressed as

$$A = \sum_{i=1}^{k} \varepsilon_{\lambda i} c_i L_i \qquad (1.7)$$

Where, k represents the number of independent absorbers.

Since arterial blood flow is pulsatile, the absorbance due to it will also be a pulsatile signal. The time period of each pulse is dictated by the heartbeat and the amplitude by the concentration of various constituent parts of arterial blood and path length of light traveling through the arteries. In human blood (solution), the main light absorbers (solutes) are the hemoglobin. Previous research had indicated that oxy and deoxy-hemoglobin have different optical attenuation characteristics [32]-[34] as given in Fig. 1.3. For best results, the wavelengths are to be selected such that at one wavelength, the attenuations by Hb and HbO are as different as possible and at the second wavelength they are nearly the same (Optimal: be 600 nm & 798 nm). Practically, the wavelengths are chosen based on available resources. Currently available pulse oximeters use light at 660 nm (red) and 940 nm (IR) wavelengths. Either the transmitted light through an extremity such as finger tip or ear lobe or the reflected light at these wavelengths are detected to obtain two PPG signals, say PPG_R and PPG_{IR}. The DC values and AC amplitudes of the cardiac synchronous pulsatile portions in the red and IR PPG signals, DC_R and DC_{IR} and AC_R and AC_{IR} respectively are extracted. Then a normalized red to IR absorption ratio R is obtained as

$$R = \frac{AC_R/DC_R}{AC_{IR}/DC_{IR}} \qquad (1.8)$$

Data are obtained from a group of healthy nonsmoking young volunteers made to breathe various hypoxic gas mixtures to regulate their arterial oxygen saturation

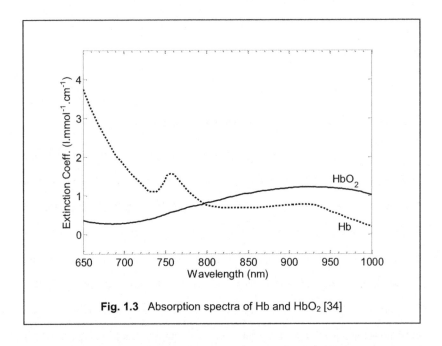

Fig. 1.3 Absorption spectra of Hb and HbO$_2$ [34]

between 70 % and 100 % [35], [36]. Samples of their arterial blood are drawn at regular intervals and the oxygen saturation (SaO$_2$) values are measured using *in-vitro* laboratory CO-Oximeter. At the same time, normalized ratios (R) are calculated from the red and IR PPG signals obtained from the volunteers. The values obtained by CO-Oximeter and R are then plotted to form a calibration curve. A typical calibration curve used by Ohmeda pulse oximeter [37] is shown in Fig. 1.4. An empirical linear approximation to the calibration curve [9], [38] is given as

$$SpO_2 = (110 - 25R)\% \qquad (1.9)$$

Each pulse oximeter manufacturer will have their own calibration curve(s) based on the group tested and in most cases, an $R = 1$ approximately indicates a saturation of 85 %. In standard pulse oximeter algorithm, once the R is calculated from the two PPG signals, SpO$_2$ values are determined from R using a look-up

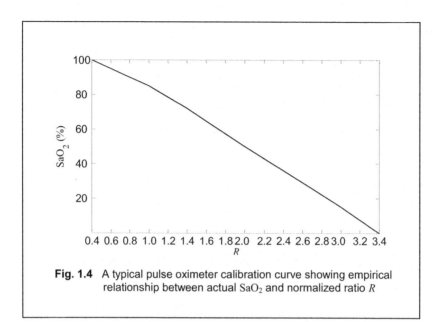

Fig. 1.4 A typical pulse oximeter calibration curve showing empirical relationship between actual SaO_2 and normalized ratio R

table corresponding to the empirically derived calibration curve. A microcontroller sitting inside the pulse oximeter performs the calculation of R from the acquired PPG signals and displays corresponding SpO_2 from a look-up table.

1.7 Problems Associated with Pulse Oximetry

Pulse oximetry has made a significant contribution to the noninvasive monitoring of not only oxygen saturation in arterial blood but also heart rate and heart rate variability in a wide variety of clinical situations and allows for continuous, quick and reliable measurement of oxygen saturation while avoiding the discomfort and risks of arterial puncture. However, present day pulse oximeters have shortcomings and research is being carried out all over the globe in removing these shortcomings. The factors adversely affecting the accuracy of pulse oximeter output include probe placement, peripheral vasoconstriction (poor perfusion) [39], dyshemoglobins [40], intravascular dyes [41], ambient light [42], nail polish [43], [44], skin pigmentation [45], and reliability problems due to movement of patient [46], [47]. Some limitations of pulse oximetry are attributable to the design of pulse oximeters and the principle of measurement employed. As pulse oximeters only use two wavelengths, they can measure only two hemoglobins and thus, estimate functional saturation, not fractional saturation. Other compounds that absorb light at the same wavelengths will thus introduce errors. Important limitation of pulse oximetry is the presence of abnormal hemoglobins. The most common and potentially most serious is carboxyhemoglobin, which the oximeter detects as oxyhemoglobin and thus overestimates the true concentration of oxyhemoglobin. Other dyshemoglobins such as methemoglobin also interfere [48], [49]. CO-Oximetry should be employed if the presence of abnormal hemoglobin is expected.

Pulse oximetry measurements are also affected by various dyes/ pigments. For example, *methylene blue*, which is recommended as a therapy for methemoglobinemia, absorbs light at 660 nm, similar to the absorption rate of Hb and hence the presence of methylene blue will make a pulse oximeter read erroneously [50]. Other intravenous dyes (e.g., fluorescein, indocyanine green, and indigo carmine) used for therapeutic or diagnostic purposes also will produce spuriously low pulse oximeter readings [51].

Pulse oximeters estimate SpO_2 level by computing the ratio of pulsatile to baseline tissue absorption, thus, an adequate arterial pulsation is required. A significant decrease in peripheral vascular pulsation (such as in shock, severe hypotension, vasoconstriction or hypothermia), will produce unreliable oximeter readings [52] as the detected signal can not be differentiated from background noise. Though most pulse oximeters are designed to display a message such as inadequate pulse signal (low perfusion), it often turns out that a measurement can not be made during such complications. The only option then is to change the location of the sensor.

Excessive light from number of sources including surgical lamps, infrared, xenon and fluorescent lamps can interfere with oximetry accuracy [53]. In ambient light interference the end result is addition of false signals to both the red and IR PPG signals. However, a recent study reveled that ambient light has no statistically significant effect on pulse oximetry readings [54]. Even had the differences been statistically significant, the magnitude of the differences was small and thus clinically unimportant. Interference with high intensity ambient light can be easily avoided by placing an opaque material (dark colored cloth) to surround the probe. Studies on the effect of skin pigmentation on the pulse oximetry revealed that dark skin color may influence the performance and concluded that accuracy was slightly less for subjects with dark skin color than those with lighter skin color [45].

Fingernail polish has been cited as a variable affecting pulse oximetry readings when a finger or toe probe is used. Black, blue, and green nail polish significantly lowered oximeter readings of oxygen saturation [43], [55]. A nursing recommendation is to remove fingernail polish or artificial nails before routine pulse oximetry monitoring. However, a recent study [44] documented a decrease (2%) in pulse oximetry readings occurring only with brown or black fingernail polish. The effect due to nail polish could be eliminated by placing the finger probe side to side instead of top to bottom.

In a pulse oximeter clean artifact-free PPG signals with clearly separable DC and AC parts are necessary for error-free SpO_2 estimation. Various situations produce low signal-to-noise ratio on the detected PPG signals leading to errors in the measured values of SpO_2. Mostly inaccurate readings in a pulse oximeter arise due to the artifacts created in the detected PPG signals by the movement of a patient. Since the pulsatile component is quite small in a PPG (0.1% of total signal amplitude) even a slightest movement of the patient will disturb the detected pulsatile component with additional pulsations other than those expected from the heart beat. Additional pulsations due to motion artifact lead to inaccurate estimation of SpO_2. The effect of motion artifact can be reduced by suitable processing of the PPG signals. To some extent, effect of motion artifact can be reduced by displaying the average value of several SpO_2 readings [38].

A straight forward solution is to altogether avoid motion artifacts by securing the sensor head rigidly to the skin of the patient at a monitoring site and avoid relative motion between the sensor and the patient. However, this solution is not practical because pressure on the skin would lead to discomfort to the patient. Hence the probe in a pulse oximeter is always designed to exert bare minimum pressure on the skin, just adequate to keep the probe in place. Moreover when pressure is applied vasoconstriction occurs resulting in reduced perfusion. Sustained pressure elevates skin temperature resulting in vasodilatation leading to increased perfusion. In both cases sweating takes place resulting in deterioration of contact between the sensor and the skin. If proper compensation is not applied for these changes, then SpO_2 readings obtained will be erroneous.

1.8 Limitations of Pulse Oximetry in Vogue and Motivation for further Research

Thus limitations of the present day pulse oximeters can be summarized as:

(i) SpO_2 estimation in vogue relies on an empirical equation realised by linear regression of the data obtained from a group of healthy and young volunteers. Therefore, the obvious limitation is that commercial pulse oximeters are as accurate as their calibration curves are. Since these curves are from 70 % to 100 % of SpO2 values, manufacturers extrapolate their 70 to 100 % results downward in order to display SpO_2 values below 70 %. Hence currently available pulse oximeters are accurate (within 2 % error) only in the range from 70 to 100 % [35], [56], [57].

(ii) Light absorption is also dependent on pigmentation and thickness of test site. It was reported that as the skin pigmentation darkens, the pulse oximeter performance deteriorates [58]. This is mainly due to a lower signal-to-noise ratio caused by increased light absorption as the skin pigmentation darkens. It would be advantageous if the calculation of SpO2 is made independent of patient dependent parameters, such as skin pigmentation and intervening volume of tissue. However, better accuracy is expected if the computation of SpO2 is obtained analytically without resorting to the use of "calibration curves".

(iii) Furthermore, the amplitudes of detected PPG signals depend on detector sensitivities and intensities of individual sources (red and IR LEDs). It would, indeed, be attractive if estimation of SpO2 is made insensitive to the variations in the intensity of the light sources and sensitivities of the detectors employed in the PPG sensor head.

(iv) Reliability of pulse oximeter performance is profoundly affected by the movement of a patient. Motion artifacts introduced in the PPG signals due to the movement of a patient result in a significant error in the readings of pulse oximeters and hence are a common cause of oximeter failure and loss of accuracy. Reduction of motion artifacts from PPG signals for reliable estimation of SpO_2, in fact, has been a challenging problem ever since the invention of pulse oximetry.

These challenges motivated our research and the work reported in this book addresses the above issues and provides novel solutions.

1.9 Objectives and Scope

This book has two main objectives. The first of these focuses on the development of alternate SpO_2 computation strategies, which make the estimation of SpO_2 to be free from patient and sensor (source and detector) dependent parameters.

The second objective is to develop signal processing methods for the reduction of motion artifacts from PPG signals, so as to improve the reliability of SpO_2 estimation employing pulse oximeters.

1.10 Organization of the book

This text is arranged into six chapters. Chapter 1 provides introduction to the physiology of oxygen transport in blood, oxygen saturation of arterial blood, photoplethysmography, principles of pulse oximetry, motivation and objectives.

Chapter 2 presents a novel model based method for the computation of SpO_2 utilizing the traditional negative feedback compensation scheme that normalizes the red and IR PPG signals. A prototype pulse oximeter that employs a sensor housing red and infrared (IR) LEDs and suitable photo detectors is developed to validate the proposed method. The developed prototype employs the audio channel of a PC for data acquisition dispensing with expensive analog to digital converter hardware. The proposed pulse oximeter estimates SpO_2 using an analytical expression and hence does not rely on calibration curves.

In Chapter 3, a novel method of estimation of SpO_2 wherein the red and IR PPG signals are processed appropriately with a view to remove the source and detector dependent parameters is presented. Appropriate expressions are derived based on a model of light propagation through tissue, bone and blood such that the equation for the computation of SpO_2, utilizing the processed PPG signals and the well-known extinction coefficients of HbO_2 and Hb, is devoid of all interfering parameters, thus obviating the need for extensive calibration.

Chapter 4 deals with a refined method of measurement of SpO_2 employing only the peak-to-peak values of the suitably processed red and IR PPG signals. This method too does not require extensive calibration needed in present day pulse oximeters that employ empirical curve fitting.

Chapter 5 starts with a brief review of existing motion artifact reduction methods applied to PPG signals and presents a method to reduce the motion artifacts from the corrupted PPG signals utilizing singular value decomposition (SVD). Application of SVD technique leads to a stable and reliable estimation of SpO_2 even when the PPG signals are distorted by motion artifacts.

Chapter 6 proposes a simple but effective method for reduction of motion artifacts using the well known Fourier series analysis (FSA) applied to a PPG signal on a cycle-by-cycle basis. Over and above artifact reduction, the FSA also provides data compression. Practical applicability of all the techniques outlined in this book is verified by experiments conducted using suitable prototype units.

2. Novel Model Based Method of Computation of SpO$_2$

2.1 Existing Methods of Pulse Oximetry - Problems and Challenges

As brought out earlier, pulse oximetry first introduced by Takuo Aoyagi *et al.*, utilizes two PPG signals, one obtained using a source possessing a wavelength λ in the red region (λ ≈ 600 nm to 700 nm) and the other in the infrared region (λ ≈ 850 nm to 1000 nm) to evaluate SpO$_2$ [27]. Either on Automatic Gain Control (AGC) scheme or a negative feedback compensation technique is traditionally employed for normalizing the red and IR PPG signals. In the negative feedback schemes, the intensities of the red and IR sources are continually adjusted to compensate for the variations in a patient's skin color and intervening volume of tissue between the sources and the detectors. The values of red and IR PPG signals at systole and diastole are traditionally used for oxygen saturation measurements in a typical pulse oximeter [9]. As already seen in the previous chapter, pulse oximeters compute the oxygen saturation by means of a linear interpolation obtained through calibration [9], [37], [38]. A novel method of computation of SpO2 using the normalized red and IR PPG signals and a model without resorting to ratio of ratios is presented in this chapter. An analytical expression derived for the computation of SpO$_2$ utilizing the normalized red and IR PPG signals and the well known extinction coefficients [34] of hemoglobin and oxyhemoglobin obviates the necessity of applying calibration curves for the determination of SpO$_2$. Thus the method presented here, not only removes the influence of patient dependent parameters, because of the well known feedback technique, in the determination of SpO_2, but also renders the pulse oximeter not to rely on calibration curves [59].

2.2 Feedback Compensation for Normalization of a PPG Signal

To obtain a PPG, a sensor housing a suitable light source to illuminate the part of body of interest and a photo detector to detect the transmitted or reflected light is necessary. If the PPG is obtained from the reflected light, the source and the detector have to be housed on the same plane as indicated in Fig. 2.1(a). Earlier research has identified that the optimum distance between the source and detector in the sensor of Fig. 2.1(a) is in the range of 4 to 5 mm [18], [60]. This type of sensor can be used over the skin on any part of the body [61], [62]. On the other hand, if one has to obtain a PPG using the transmitted light, the source and the detector need to be arranged on two different but parallel planes, as indicated in Fig.2.1 (b). It is obvious that sensors of transmission type can be applied only to extremities of the body such as the earlobes and fingertips. The fingertip is a convenient option because a good PPG signal can be obtained with least discomfort to a patient and hence is popular with most pulse oximeter sensors. Though the following explanations and equations are for PPG signals obtained using sensors of the transmission type, they are equally applicable for PPG signals obtained utilizing the reflected light as well. One only needs to replace the word "absorption" to "reflection" to obtain explanations and relevant expressions applicable to PPG signals obtained employing the reflectance type sensors. To obtain a transmission type PPG, a section of the tissue is illuminated as indicated in Fig. 2.1(b). The epidermis, skin tissue, blood, tissue and soft bone (if applicable) absorb the incident light depending on their optical characteristics.

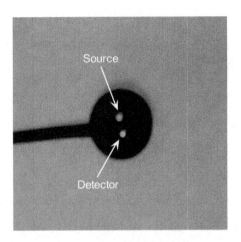

Fig. 2.1 (a) Reflectance type PPG sensor head

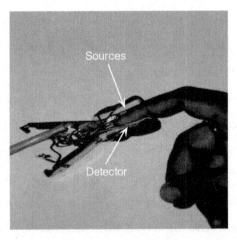

Fig. 2.1 (b) Transmission type PPG sensor head

The remaining unabsorbed light transmitted to the detector gets converted into an equivalent current or voltage by the detector resulting in a PPG, a typical form of which is illustrated in Fig. 1.2. The different waveforms in Fig. 1.2 are not drawn to scale as the magnitudes of individual components differ considerably. The DC component ($\approx$ 99 %) of the detected signal traverses through skin, tissue and bone alone and hence is dependent on the color of the skin and the volume of intervening tissue and bones. The pulsatile component ($\approx$ 0.1 % of the PPG) at the heart rate, in addition to the attenuation by epidermis, bone and tissue, is due to the attenuation by the intervening volume of arterial blood. Since the detected signal $v_{D\lambda}$ is a function of light absorption across the section of the finger being illuminated, it is a function of the concentration and attenuation characteristics of the constituent parts encountered by the light beam and hence can be represented as

$$v_{D\lambda} = K_{D\lambda}\left(I_{IN\lambda}\varepsilon_{ED\lambda}\langle ED\rangle\left(\langle TI\rangle\varepsilon_{TI\lambda} + \langle BO\rangle\varepsilon_{BO\lambda} + \langle BL\rangle\varepsilon_{BL\lambda}(\langle\alpha\rangle\varepsilon_{BO\lambda} + \langle\beta\rangle\varepsilon_{TI\lambda} + \langle\gamma\rangle\varepsilon_{TI\lambda}\varepsilon_{BO\lambda})\right)\right),$$

(2.1)

where $I_{IN\lambda}$ is the intensity of the incident light at wavelength λ, $K_{D\lambda}$ the sensitivity of the detector, $\varepsilon_{ED\lambda}$, $\varepsilon_{BO\lambda}$, $\varepsilon_{BL\lambda}$ and $\varepsilon_{TI\lambda}$ are the extinction coefficients of epidermis, bone, blood and tissue respectively at the wavelength λ. $\langle ED\rangle, \langle BO\rangle, \langle TI\rangle$ and $\langle BL\rangle$ are the cell concentrations of epidermis, bone, tissue and blood correspondingly. $\langle\alpha\rangle, \langle\beta\rangle$ and $\langle\gamma\rangle$ are equivalent concentrations of bone, tissue or both bone and tissue respectively either preceding the blood vessels or succeeding them. Extinction coefficient is a numeric measure of the opaqueness of a particular type of cell exposed to the light [63]. The greater the extinction coefficient, the greater is the opaqueness. The constants $\varepsilon_{ED\lambda}$, $\varepsilon_{BO\lambda}$, $\varepsilon_{TI\lambda}$ and

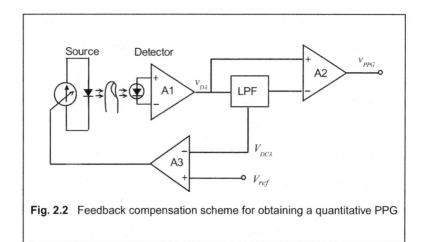

Fig. 2.2 Feedback compensation scheme for obtaining a quantitative PPG

the cell concentrations $\langle ED \rangle, \langle BO \rangle$ and $\langle TI \rangle$ are patient dependent and hence for any meaningful measurement, these must be taken into account.

Blazek *et al.* pioneered the concept of obtaining a quantitative PPG signal by employing negative feedback compensation [18]. However, in their method the feedback scheme is employed with a sample and hold inserted into the feedback loop and the feedback is enabled only for a set period just prior to a measurement. During the measurement period, the feedback is broken, albeit with the help of the sample and hold to ensure that conditions existing prior to the breaking off of the feedback are maintained during the measurement period. The feedback could not be enabled all the time in their case, since their interest is the low frequency component of a PPG signal arising due to the venous blood. In the method proposed here, a negative feedback compensation scheme without any sample and hold in the feedback path, as indicated in Fig. 2.2, is employed. In the scheme shown in Fig. 2.2, the intensity of light falling on the detector is first converted into an equivalent current by the photo diode detector. This current is converted into a voltage by the current-to-voltage converter (A1 in Fig. 2.2) is filtered by the low-pass filter (LPF) to extract the DC component. The extracted DC component is compared with a preset DC reference voltage V_{ref}. The difference between the reference and the detector output is fed back to control the intensity of the source. Let the intensity of the light be initially some value. If a patient's skin is dark in color then the light falling on the detector will be low and hence the DC part of the detector output may be lower than V_{ref}, resulting in the drive to the source to be increased suitably. On the other hand, if the patient's skin is of a lighter shade, the feedback compensation scheme will reduce the drive to the source suitably so that once again the DC output is made equal to V_{ref}. The situation is the same even if the intervening volume of tissue between the source and the detector varies. The feedback will increase or decrease the drive to the source to compensate for the variations in the thickness of the finger inserted into the sensor. Mathematically speaking, the negative feedback ensures that the DC component of the detector output $V_{DC\lambda} = K_{D\lambda}\left(I_{IN\lambda}\varepsilon_{ED\lambda}\langle ED\rangle\left(\langle TI\rangle\varepsilon_{TI\lambda}+\langle BO\rangle\varepsilon_{BO\lambda}\right)\right)$ for any patient is made equal to the pre-fixed DC reference voltage, V_{ref}, thus normalizing the PPG across different patients. Thus

$$V_{DC\lambda} = K_{D\lambda}\left(I_{IN\lambda}\varepsilon_{ED\lambda}\langle ED\rangle\left(\langle TI\rangle\varepsilon_{TI\lambda}+\langle BO\rangle\varepsilon_{BO\lambda}\right)\right) = V_{ref} \qquad (2.2)$$

Utilizing equation (2.2) and the fact that the pulsatile component of a PPG is one thousandth of its DC component, equation (2.1) can be simplified as

$$v_{D\lambda} = V_{ref}\left(1+\langle BL\rangle\varepsilon_{BL\lambda}\right) \qquad (2.3)$$

In such a situation, computation of SpO_2 from the resultant PPG can be made patient independent to a very large extent by choosing the proposed novel method of computation as shown next.

2.3 Novel Method of Computation of SpO₂ Utilizing the Normalized PPG Signals

The time varying pulsatile component $v_\lambda = V_{ref}(\langle BL \rangle \varepsilon_{BL\lambda})$ in equation (2.3) is due to the arterial blood that contains 55 % plasma, 43 % red blood cells (called erythrocytes, made of hemoglobin molecules), 1.5 % of white blood cells (leukocytes) and 0.5 % platelets (thrombocytes). Since the amount of white blood cells and platelets are negligible compared to the red blood cells, they have insignificant influence on the absorption of light. As the plasma has nearly zero optical attenuation in the wavelength region of interest [64], we need to consider only the attenuation due to red blood cells. Moreover, red blood cells predominate over the other blood components by two to three orders of magnitude with regard to absorption [65]. The red blood cells are made of hemoglobin molecules some carrying oxygen molecules (HbO₂) and the rest without oxygen molecules attached to them (Hb). Utilizing these facts, equation (2.3) is modified as

$$v_\lambda \approx V_{ref}\left(\langle Hb \rangle \varepsilon_{Hb\lambda} + \varepsilon_{HbO\lambda}\langle HbO_2 \rangle\right)$$

Here $\varepsilon_{Hb\lambda}$ and $\varepsilon_{HbO\lambda}$ are the extinction coefficients of hemoglobin and oxy-hemoglobin at the wavelength λ. On simplification the pulsatile portion becomes

$$v_\lambda \approx V_{ref}\left(\varepsilon_{Hb\lambda} + \varepsilon_{HbO\lambda}\frac{\langle HbO_2 \rangle}{\langle Hb \rangle}\right)\langle Hb \rangle = V_{ref}\left(\varepsilon_{Hb\lambda} + \varepsilon_{HbO\lambda}Q\right)\langle Hb \rangle \qquad (2.4)$$

The absorption spectra of Hb and HbO₂ are well documented [32]-[34] and is reproduced in Fig. 2.3. A marked difference in the absorption characteristics of Hb and HbO₂ can be seen at wavelengths 660 nm and 940 nm. The wavelength at which Hb and HbO₂ absorb to the same extent is called isobestic point. From Fig. 2.3, it is seen that the isobestic point of Hb and HbO₂ occurs at 795±1.5 nm.

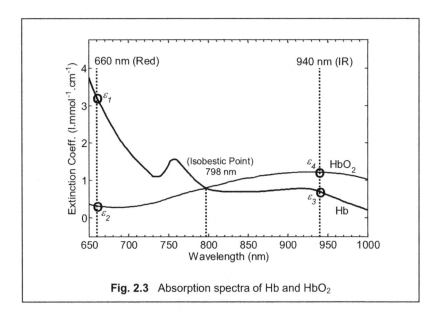

Fig. 2.3 Absorption spectra of Hb and HbO$_2$

These points may be used as reference points where light absorption is independent of the degree of saturation. The peak-to-peak value $V_{P\lambda}$ of the pulsatile signal v_λ is

$$V_{P\lambda} \approx V_{ref}\left(\varepsilon_{Hb\lambda} + \varepsilon_{HbO\lambda}Q\right) \tag{2.5}$$

The peak-to-peak values V_{PR} and V_{PIR} of the pulsatile portions of the outputs v_R and v_{IR} at red and IR wavelengths respectively are

$$V_{PR} \approx V_{ref}\left(\varepsilon_1 + \varepsilon_2 Q\right) \tag{2.6}$$

$$V_{PIR} \approx V_{ref}\left(\varepsilon_3 + \varepsilon_4 Q\right) \tag{2.7}$$

Here ε_1 and ε_2 are the extinction coefficients of Hb and HbO$_2$ respectively at red wavelength. ε_3 and ε_4 are the extinction coefficients of Hb and HbO$_2$ respectively at infrared wavelength, obtained from Fig. 1.3, as indicated in Fig. 2.3. Dividing equation (2.6) by equation (2.7) results in,

$$\frac{V_{PR}}{V_{PIR}} = \frac{(\varepsilon_1 + \varepsilon_2 Q)}{(\varepsilon_3 + \varepsilon_4 Q)} \qquad (2.8)$$

Solving for Q from equation (2.8)

$$Q = \frac{(V_{PIR}\varepsilon_1 - V_{PR}\varepsilon_3)}{(V_{PR}\varepsilon_4 - V_{PIR}\varepsilon_2)} \qquad (2.9)$$

From equations (2.9) and (1.3), we get:

$$SpO_2 = \frac{(V_{PIR}\varepsilon_1 - V_{PR}\varepsilon_3)}{(V_{PIR}\varepsilon_1 + V_{PR}\varepsilon_4) - (V_{PIR}\varepsilon_2 + V_{PR}\varepsilon_3)} 100\% \qquad (2.10)$$

It is seen that none of the terms in equation (2.10) is patient and sensor dependent and hence the equation is free of any patient and sensor dependent parameters. To validate the above proposed scheme, a prototype detailed in the sequel is built and tested and the inherent error in the proposed measurement is dealt with next.

2.4 Possible Sources of Errors

In deriving equations (2.8) and (2.9), it is assumed that the individual feedback compensation schemes of the red and IR channels are identical and hence the ratio of the DC components of the red and IR PPG signals, say η is taken as 1. In a practical situation, there may be a small deviation and η may not be exactly 1. In that case

$$SpO_2 = \frac{(\eta V_{PIR}\varepsilon_1 - V_{PR}\varepsilon_3)}{(\eta V_{PIR}\varepsilon_1 + V_{PR}\varepsilon_4) - (\eta V_{PIR}\varepsilon_2 + V_{PR}\varepsilon_3)} 100\% \qquad (2.11)$$

The error due to mismatch in the DC component of red and IR PPG signals is made insignificant if the same, precision and stable V_{ref} is employed for the individual feedback loop and a sufficiently high loop gain is chosen to reduce static errors in the feedback loop. Table 2.1 provides possible worst case error for

Table 2.1 Error in the computation of SpO$_2$ due to mismatch in the DC voltages of red and IR PPG signals.

Sl. No	η	Computed SpO$_2$ including η	Error (%)
1	0.97	98.6065	-0.3595
2	0.98	98.7274	-0.2373
3	0.99	98.8460	-0.1175
4	1.01	99.0764	0.1153
5	1.02	99.1883	0.2284
6	1.03	99.2982	0.3394

Actual SpO$_2$ is 98.9623

various values of η. However in a practical case, η will be in the range 0.9999 to 1.0001. Hence the error due to mismatch in DC voltages of red and IR PPG signals is negligible in a practical case. The wavelengths of the LEDs used must be known exactly to extract the correct values of ε_1, ε_2, ε_3 and ε_4. Present day LEDs produce a narrow band of wavelengths with a rated center wavelength. The variation in the expected wavelengths [66] of red and IR result in the errors δ_1, δ_2, δ_3 and δ_4 in ε_1, ε_2, ε_3 and ε_4 respectively. The errors δ_{VR} and δ_{VIR} in the measurement of V_{PR} and V_{PIR} and errors δ_1, δ_2, δ_3, and δ_4 will result in δ_{SpO2}, which is the error in the determination of SpO$_2$ as

$$\delta_{SpO2} = \frac{a}{b-a}(\delta a - \delta b) \tag{2.12}$$

Where $a = V_{PR}\varepsilon_4 - V_{PIR}\varepsilon_2$, $b = V_{PR}\varepsilon_3 - V_{PIR}\varepsilon_1$

$$\delta a = \left(V_{PR}\varepsilon_4(\delta_{VR}+\delta_4) - V_{PIR}\varepsilon_2(\delta_{VIR}+\delta_2)\right)/a$$

$$\delta b = \left(V_{PR}\varepsilon_3(\delta_{VR}+\delta_3) - V_{PIR}\varepsilon_1(\delta_{VIR}+\delta_1)\right)/b$$

Table 2.2 Error in the computation of SpO_2 due to errors in choosing extinction coefficients.

Sl. No	LEDs Wavelengths (nm)	Errors in extinction coefficients (%)	Estimated error in SpO_2 (%)
1	658, 930	δ_1 = -2.8203, δ_2 = -1.8428 δ_3 = -2.6498, δ_4 = -3.6007	0.1930
2	658, 950	δ_1 = -2.8203, δ_2 = -1.8428 δ_3 = 23.472, δ_4 = -2.1595	0.0100
3	664, 930	δ_1 = 5.6503, δ_2 = 3.6316 δ_3 = -2.6498, δ_4 = -3.6007	0.8636
4	664, 950	δ_1 = 5.6503, δ_2 = 3.6316 δ_3 = 23.472, δ_4 = -2.1595	0.6869

Range of wavelengths of LEDs: (658 nm to 664 nm) and (930 nm to 950 nm)

Of these, δ_{VR} and δ_{VIR} are dictated by the measurement set-up. For the prototype, a 16-bit analog to digital converter (ADC) with a measurement error of 0.01 % is employed, thus making the error due to V_{PR} and V_{PIR} insignificant. The values of δ_1, δ_2, δ_3, and δ_4 are dependent on the purity and stability of the outputs of the red and IR LEDs. For the prototype, the 660 nm and 940 nm LEDs used have a rated wavelength variation of 658 nm – 664 nm in the red region and 930 nm – 950 nm in the IR region. Even though δ_3 is 23.472%, its influence on computation of SpO_2 is minimal. The estimated worst case errors in four possible sets of extreme wavelengths are presented in Table 2.2 resulting in an overall error of less than 1%. This error is the only significant error and care must be taken to keep this error, a minimum by choosing the correct values of ε_1, ε_2, ε_3 and ε_4

corresponding to the emitted wavelengths of the red and IR LEDs employed. Even if the exact wavelengths of the outputs are not known, it is possible to compute the worst case error and compensate the same by calibration.

2.5 Experimental Setup and Results

A clip-on finger sensor, shown in Fig. 2.1(b), is fabricated with a red LED providing light at 660 nm wavelength and an IR LED with a light output at 940 nm serving as sources and a photodiode as detector. The LEDs and the photo diode are connected to an electronic signal conditioning circuit, shown in Fig. 2.4, for obtaining two PPG signals, one at the red wavelength and the other at the IR. To avoid interference between the red and IR lights, the individual drives to the LEDs are time sliced (when red LED is ON, the IR is kept OFF and vice versa) at a rate of 1 kHz. Due to the multiplexing of the red and IR LEDs, the output of the photodiode will contain both the red and IR PPG signals multiplexed at 1 kHz, and hence the photodiode output should be de-multiplexed. The circuit shown in Fig. 2.4 accomplishes this task as well.

Whenever clock is high all the SPDT switches (S1, S2 and S3) are set to position 1. Opamp OA1 in conjunction with transistor TR1 and the current setting resistor R_{SR} sets the current I_R (=V_{ref}/R_{SR}) through the red LED (shown as sensor LED D2 in Fig. 2.4) kept on the clip-on sensor. When the clock turns low, all the switches are changed to position 2. The current I_R now flows through the dummy red LED D1 and the sensor LED D2 is switched off. When switch S2 is in position 2, current through IR LED D4 kept on the sensor, is set as I_{IR} by opamp OA2 in conjunction with transistor TR2 and resistor R_{SIR}. Whenever clock is high this current is sent through the dummy IR LED D3, thus switching off the IR LED D4 of the sensor. When clock is high, switch S3 is in poison 1 and hence

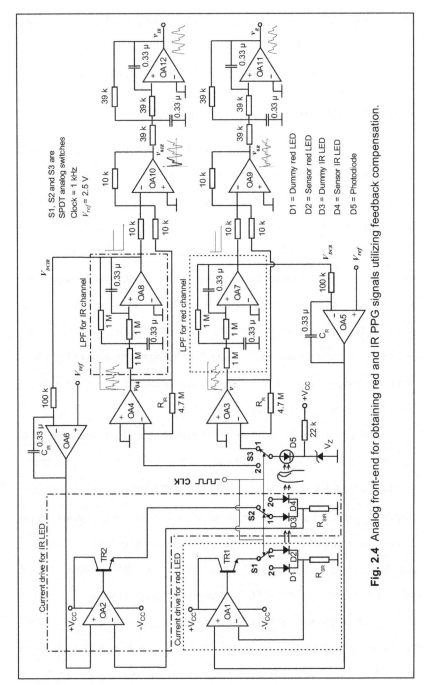

Fig. 2.4 Analog front-end for obtaining red and IR PPG signals utilizing feedback compensation.

photodiode D5 is connected to the *i-to-v* converter consisting of opamp OA3 and resistor R_R.

The red light from D2 transmitted through the finger is converted to a proportional photoelectric current by the photodiode. This current flows through R_R and produces a proportional voltage v_{03} at the output of opamp OA3. The output of OA3 is fed as input to a second order low-pass filter (LPF) with a cut-off frequency of 0.1 Hz, realized with opamp OA7. The output of the LPF, the DC part of v_{03}, is compared with the reference voltage V_{ref} and the difference controls OA1, setting the current I_R. The capacitor C_R across OA5 introduces a dominant pole and ensures a stable negative feedback system. The DC part of the output of OA3, V_{DCR} obtained from the output of OA7, is subtracted from v_{03} with the help of an inverting summer realized using opamp OA9. Hence the output of OA9 will be the pulsatile portion v_{0R} of the red PPG. v_{0R} is sent through an LPF (OA11) with a cutoff frequency of 20 Hz to remove interference at power supply frequency and red PPG v_R is obtained at the output of OA11. Similarly, the circuit part with OA2, OA4, OA6, OA8, OA10 and OA12 provide the pulsatile portion of the IR PPG v_{IR} at the output of OA12. All the opamps are type OP07 and the switches are CD4053. The 1 kHz clock is derived using an LM555 working as an astable multivibrator at 2 kHz feeding a divide-by-two flip-flop. For ease of understanding, typical waveforms at cardinal points are indicated in Fig. 2.4 itself. The signals v_R and v_{IR}, thus obtained are interfaced to a PC utilizing the audio channel, thus dispensing with expensive analog to digital hardware. Since the frequency response of the audio channel is not suitable for the red and IR PPG signals, the frequencies of the PPG signals are shifted to 2 kHz utilizing frequency modulation. The modulators were built around a couple of ADVFC32. The center frequencies of both the FM blocks were kept at 2 kHz and the

Table 2.3 Extinction coefficients employed in the prototype

Wavelength (nm)	Hb	HbO$_2$
660	$\varepsilon_1 = 3.226$	$\varepsilon_2 = 0.319$
940	$\varepsilon_3 = 0.693$	$\varepsilon_4 = 1.214$

frequency deviation chosen as 200 Hz. A suitable program was developed under the LabVIEW environment to

(i) Acquire the FM modulated PPG data from the sound card channels

(ii) Demodulate the acquired signals and obtain samples of v_R and v_{IR}

(iii) Compute and display the pulse rate by using the time period of each cycle of the IR PPG

(iv) Compute and display SpO$_2$ as given by equation (2.10)

The values of the extinction coefficients (l.mmol^{-1}.cm^{-1}) employed for the computations in the prototype are summarized in Table 2.3. A compensation factor of four is used to account for the four hemes per hemoglobin molecule [42]. Photograph of the printed circuit board of the prototype circuit given in Fig. 2.4 is shown in Fig. 2.5 (a) and the photograph of the complete prototype is shown in Fig. 2.5(b). Fig. 2.6 shows the snapshot of the front panel of the prototype virtual instrument, developed under LabVIEW environment, during a typical test. The program developed is listed in Annexure 1. Tests were conducted on 31 volunteers and the results are indicated in Table 2.4. The SpO$_2$ readings from the developed instrument were compared with the ones obtained by a commercial pulse oximeter (CPO), model Planet 50, manufactured by Larson & Toubro Limited and the readings of the prototype were within ± 2 % of the CPO.

Fig. 2.5(a) PCB of the analog signal processing part for processing PPG signals

Fig. 2.5(b) The developed photoplethysmograph for pulse oximetry

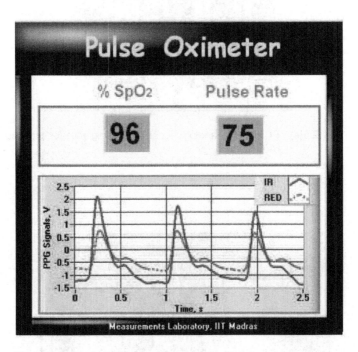

Fig. 2.6 Snapshot of the front panel of the developed instrument

Table 2.4 Comparison of SpO$_2$ values obtained with the prototype and a CPO

Subject no.	Average SpO$_2$ (Std. Dev)	
	Proposed method	CPO
1 (D,M)	97.26 (0.52)	98.85 (0.26)
2 (F,M)	97.74 (0.61)	98.81 (0.35)
3 (F,M)	98.58 (0.68)	99.42 (0.24)
4 (N,M)	97.80 (0.55)	98.22 (0.36)
5 (N,Tn)	98.96 (0.71)	98.57 (0.33)
6 (F,Tk)	97.32 (0.55)	98.55 (0.41)
7 (D,Tk)	96.86 (0.68)	97.93 (0.46)
8 (N,Tn)	97.59 (0.72)	98.19 (0.28)
9 (F,Tk)	97.97 (0.53)	98.83 (0.33)
10 (F,M)	97.38 (0.64)	98.76 (0.42)
11 (N,Tk)	96.96 (0.69)	98.25 (0.38)
12 (N,M)	97.44 (0.65)	98.83 (0.44)
13 (N,Tk)	98.06 (0.72)	98.99 (0.37)
14 (N,Tk)	98.72 (0.76)	99.22 (0.36)
15 (D,Tn)	97.26 (0.68)	98.47 (0.31)
16 (N,M)	97.49 (0.57)	98.66 (0.28)
17 (N,M)	98.27 (0.71)	99.14 (0.29)
18 (N,M)	98.05 (0.83)	98.29 (0.28)
19 (D,M)	97.19 (0.66)	98.55 (0.43)
20 (F,Tk)	96.99 (0.81)	97.82 (0.50)
21 (F,Tn)	97.38 (0.53)	98.61 (0.24)
22 (D,M)	97.77 (0.54)	98.59 (0.49)
23 (N,Tn)	97.32 (0.71)	98.70 (0.44)
24 (N,M)	97.91 (0.52)	99.13 (0.61)
25 (F,M)	96.53 (0.76)	97.81 (0.45)
26 (F,Tn)	96.85 (0.72)	98.22 (0.31)
27 (N,M)	98.09 (0.41)	99.18 (0.34)
28 (N,M)	96.76 (0.50)	98.33 (0.38)
29 (D,Tn)	97.89 (0.66)	98.29 (0.28)
30 (N,M)	98.36 (0.75)	99.55 (0.36)
31 (N,M)	98.22 (0.83)	99.13 (0.47)

Terms within brackets indicate skin color and finger thickness of subject respectively.
Skin color: **Fair**(F), **Dark**(D), **Normal**(N)
Finger thikness: **Thick**(Tk), **Thin**(Tn), **Medium**(M)

Table 2.5 SpO$_2$ computed using proposed method on L&T data				
S. No	SpO$_2$ of L&T Data	SpO$_2$ estimated from proposed method	Error (%)
1	100	98	-2.00
2	98	96	-2.04
3	96	94	-2.08
4	94	92	-2.12
5	92	91	-1.08
6	90	89	-1.11
7	85	86	+1.18
8	80	84	+5.00

Due to lack of clinical facilities in the institute, volunteers were not given a mixture of hypoxic gases so as to lower their oxygen level in arterial blood. However, upon request, L&T Medical Systems, Mysore, India, provided red and IR PPG data for SpO$_2$ ranging from 80 to 100%. The proposed model based method without resorting to the ratio of ratios technique was applied on this data and the estimated SpO$_2$ values rounded-off to nearest integer are listed in Table 2.5. The results in Table 2.5 indicate that down to 85 % of SpO$_2$, errors in the proposed method are within the acceptable limit of ± 2 %. The excessive error of 5 % at 80 % SpO$_2$ needs further validation.

A new technique of pulse oximetry is presented in this chapter. Traditional feedback compensation is employed to remove the influence of patient and sensor dependent parameters and obtain quantitative red and IR PPG signals. The novel method of computation of SpO$_2$ presented here, utilizing the quantitative red and IR PPG signals, obviates the need for calibration curves. Though the proposed method provides patient independent and calibration curve independent estimation of SpO$_2$, the feedback employed makes the oximeter sluggish. Hence, alternative methods that are based on a better model of light absorption through the finger are developed and are as detailed in the sequel.

3. A Novel Slope Based Method of Measurement of SpO$_2$

In the previous chapter, a novel method of pulse oximetry based on a bulk model is presented. The traditional negative feedback compensates for the variations in the color of the skin and the intervening volume of the tissue of test subjects. The model given in Chapter 2 is a first step in obtaining a method of computation of SpO$_2$ without resorting to *ratio of ratios* and calibration curves. A method based on a more detailed model that does not require feedback compensation to normalize the PPG signals is proposed in this chapter. First, a model for the light propagation through test object (finger) is presented. Based on the model, an appropriate signal processing method is devised to extract patient and sensor independent pulsatile components from the red and IR PPG signals that represent the blood volume changes alone. An expression for calculation of SpO$_2$, utilizing the slopes and peak-to-peak values of the processed PPG signals to compute SpO$_2$ readings directly, is derived. The method presented here dispenses the need for feedback compensation and the necessity of calibration curves.

3.1 Modeling Light Propagation through Test Object (Finger)

As in the existing methods here too a sensor as portrayed in Fig. 2.1(b) with red and IR LED sources and photo diode detector is utilized to obtain the red and IR PPG signals. The incident red or IR light from the LED travels through the finger and the transmitted light after absorption within the finger is detected by the photo diode. Since the total volume covered by the blood vessels is quite small, majority of the photons emitted from the sources go through the path made of epidermis-tissue-soft bone-tissue-epidermis and reach the detector. Only a small fraction of the photons emitted by the source goes through the path that includes blood vessels [18] (epidermis-tissue–blood-tissue-bone-tissue-epidermis or epidermis-tissue-bone-tissue-blood–tissue–epidermis or epidermis-tissue–blood-tissue-bone-tissue-blood–tissue–epidermis). The DC component of the resultant PPG, as given in Fig. 1.2, is due to the photons traveling through the path epidermis-tissue-bone-tissue-epidermis, the very low frequency component is due to the photons taking the path epidermis-tissue-venous blood-bone-tissue-epidermis and the third component at the frequency of heart rate is due to the light path epidermis-tissue-arterial blood-bone-tissue-epidermis [59].

The light path through the finger is modeled as shown in Fig. 3.1. Here the beam of light is assumed to have a cylindrical shape of length L with uniform cross section of area A. A disc of thickness dl in this path is considered; the input intensity of light on the disc is taken as i_l and the attenuation across the disc is indicated as di_l. Nearly 99 % of the volume of the cylinder under consideration is made of molecules of dermis, tissue and bones. Hence the attenuations due to these cells are considered first. The cross sectional areas of dermis, tissue and

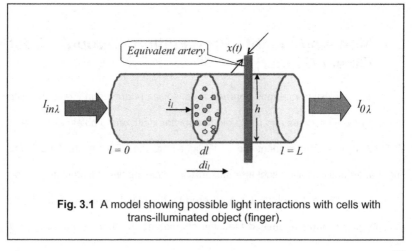

Fig. 3.1 A model showing possible light interactions with cells with trans-illuminated object (finger).

soft bone cells be A_{DE}, A_{TI} and A_{BO} respectively and their optical attenuation at a wavelength λ be $\alpha_{DE\lambda}$, $\alpha_{TI\lambda}$, and $\alpha_{BO\lambda}$ respectively. Each cell of a particular type attenuates certain amount of light passing through that cell depending on its optical property and also scatters the incoming light.

As an example, the attenuation characteristics of a tissue cell is portrayed in Fig. 3.2(a). The attenuation by the cell indicated in Fig. 3.2(a) can be considered as though a fraction of the cross sectional area of that cell is completely opaque and the rest of the cross sectional area of that cell completely transparent as indicated in Fig. 3.2(b). Then the total opaque area on the disc of area A and thickness dl due to a particular type of cell, say tissue, will be $\alpha_{TI\lambda} A_{TI} N_{TI} A\, dl$, where N_{TI} is the number of tissue cells per unit volume (cells/m³). Similarly the attenuation due to other cells can also be obtained and the total attenuation di_l

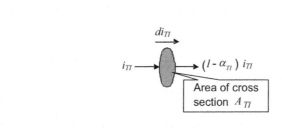

Fig. 3.2(a)   Optical attenuation through a cell (tissue)

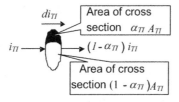

Fig. 3.2(b)   Optical attenuation through a cell – equivalent representation

across the disc of length dl is expressed as

$$di_i = -\frac{\alpha_{DE\lambda}A_{DE}N_{DE} + \alpha_{TI\lambda}A_{TI}N_{TI} + \alpha_{BO\lambda}A_{BO}N_{BO}}{A} A\, dl\, i_i \qquad (3.1)$$

The scattering effect is also included in the attenuation. Rearranging equation (3.1) we get

$$\frac{di_i}{i_i} = -(\alpha_{DE\lambda}A_{DE}N_{DE} + \alpha_{TI\lambda}A_{TI}N_{TI} + \alpha_{BO\lambda}A_{BO}N_{BO})\, dl$$

Integrating from 0 to L gives

$$\int_0^L \frac{di_i}{i_i} = -\int_0^L (\alpha_{DE\lambda}A_{DE}N_{DE} + \alpha_{TI\lambda}A_{TI}N_{TI} + \alpha_{BO\lambda}A_{BO}N_{BO})\, dl$$

Evaluating the integrals results in

$$\ln I_{0\lambda} - \ln I_{in\lambda} = -(\alpha_{DE\lambda}A_{DE}N_{DE} + \alpha_{TI\lambda}A_{TI}N_{TI} + \alpha_{BO\lambda}A_{BO}N_{BO})\, L$$

Taking $\ln^{-1}$,

$$I_{0\lambda} = I_{in\lambda} e^{-(\alpha_{DE\lambda}A_{DE}N_{DE} + \alpha_{TI\lambda}A_{TI}N_{TI} + \alpha_{BO\lambda}A_{BO}N_{BO})L}$$

Here $I_{in\lambda}$ is the input light intensity generated by the LED and $I_{0\lambda}$ is the output light intensity detected by the photo diode. Hence the DC output from the photo diode circuit is

$$V_{0DC\lambda} = K_{D\lambda}I_{0\lambda} = K_{D\lambda}I_{in\lambda} e^{-(\alpha_{DE\lambda}A_{DE}N_{DE} + \alpha_{TI\lambda}A_{TI}N_{TI} + \alpha_{BO\lambda}A_{BO}N_{BO})L} \qquad (3.2)$$

Here $K_{D\lambda}$ is the sensitivity of the photo detector circuit at a given wavelength λ. A small fraction, say $\sigma I_{0\lambda}$, of the light also travels through blood vessels. The cross sectional areas of Hb and HbO$_2$ are chosen as A_{Hb} and A_{HbO} and their optical attenuations as $\alpha_{Hb\lambda}$ and $\alpha_{HbO\lambda}$ respectively. All the individual arteries that are in the path of light are now considered as a single equivalent artery having an equivalent length of interaction with light as h and equivalent width $x(t)$ as indicated in Fig. 3.1. As blood is pumped into the arteries, the vessels enlarge and shrink at the rate set by the heart and hence $x(t)$ is a time varying quantity. If

$\hat{x}$ is the maximum width of the equivalent artery (width of the equivalent artery when the blood flowing in it, is a maximum), then the attenuation due to arterial blood $di_{0x\lambda}$ is given as

$$di_{0x\lambda} = -\sigma i_{0x\lambda} \frac{(\alpha_{Hb\lambda}A_{Hb}N_{Hb} + \alpha_{HbO\lambda}A_{HbO}N_{HbO})x(t)h}{\hat{x}h} dx \qquad (3.3)$$

Here $x(t)h$ is the total equivalent area occupied by the arteries (or equivalent artery) in the light path at a given instant of time t and $\hat{x}h$ is the maximum equivalent area of arterial blood vessels in the path of light. Rearranging equation (3.3) and integrating with limits 0 to $\hat{x}$ results in

$$\int_0^{\hat{x}} \frac{di_{0x\lambda}}{i_{0x\lambda}} = -\int_0^{\hat{x}} \frac{(\alpha_{Hb\lambda}A_{Hb}N_{Hb} + \alpha_{HbO\lambda}A_{HbO}N_{HbO})x}{\hat{x}} dx$$

$$\ln \frac{i_{0x\lambda}}{\sigma I_{0\lambda}} = -\frac{(\alpha_{Hb\lambda}A_{Hb}N_{Hb} + \alpha_{HbO\lambda}A_{HbO}N_{HbO})x^2}{2\hat{x}} \bigg|_{x=0}^{x=\hat{x}}$$

Here $i_{0x\lambda}$ is the time varying component of light intensity received at the detector and $\sigma I_{0\lambda}$ is the peak value of that portion of the light intensity attenuated by arteries also, apart from dermis, tissue and soft bones. Hence the pulsatile component at the output of the detector is

$$v_{0AC\lambda} = K_{D\lambda}i_{0x\lambda} = K_{D\lambda}\sigma I_{0\lambda} e^{-\frac{(\alpha_{Hb\lambda}A_{Hb}N_{Hb} + \alpha_{HbO\lambda}A_{HbO}N_{HbO})x^2}{2\hat{x}}\bigg|_{x=0}^{x=\hat{x}}} \qquad (3.4)$$

3.2 The Slope Based Method of Measurement of SpO$_2$

Taking the natural logarithm of the pulsatile output $v_{0AC\lambda}$ given in equation (3.4), we get

$$\ln(v_{0AC\lambda}) = \ln(K_{D\lambda}\sigma I_{0\lambda}) - \frac{(\alpha_{Hb\lambda}A_{Hb}N_{Hb} + \alpha_{HbO\lambda}A_{HbO}N_{HbO})x^2}{2\hat{x}} \bigg|_{x=0}^{x=\hat{x}} \qquad (3.5)$$

Traditionally, the optical attenuation characteristics of Hb and HbO$_2$ are given in terms of their extinction coefficients [34]. If we use the extinction coefficients, then equation (3.5) gets modified as

$$\ln(v_{0AC\lambda}) = \ln(K_{D\lambda}\sigma\, I_{0\lambda}) - \frac{(\varepsilon_{Hb\lambda}\langle Hb\rangle + \varepsilon_{HbO\lambda}\langle HbO_2\rangle)x^2}{2\hat{x}}\bigg|_{x=0}^{x=\hat{x}},$$

where $\varepsilon_{Hb\lambda}$ and $\varepsilon_{HbO\lambda}$ are the extinction coefficients of Hb and HbO$_2$ at the wavelength λ. $\langle Hb\rangle$ and $\langle HbO_2\rangle$ are the cell concentrations of Hb and HbO$_2$ in arterial blood. In the above equation, the term $\ln(K_{DR}\sigma\, I_{0R})$ is a constant for a given subject. Hence the signal due to blood volume change alone in the arteries is

$$v_{0\lambda} = \ln(v_{0AC\lambda})\big|_{pulse} = -\frac{(\varepsilon_{Hb\lambda}\langle Hb\rangle + \varepsilon_{HbO\lambda}\langle HbO_2\rangle)x^2}{2\hat{x}}\bigg|_{x=0}^{x=\hat{x}} \quad (3.6)$$

The peak-to-peak value of the pulsatile portion of $v_{0\lambda}$ in equation (3.6) is

$$V_{p\lambda} = -\frac{(\varepsilon_{Hb\lambda}\langle Hb\rangle + \varepsilon_{HbO\lambda}\langle HbO_2\rangle)\hat{x}}{2}$$

Rearranging the above equation results in

$$\hat{x} = -\frac{2V_{p\lambda}}{(\varepsilon_{Hb\lambda}\langle Hb\rangle + \varepsilon_{HbO\lambda}\langle HbO_2\rangle)}. \quad (3.7)$$

Substituting the value of $\hat{x}$ from equation (3.7) in equation (3.6);

$$v_{0\lambda} = \frac{(\varepsilon_{Hb\lambda} + \varepsilon_{HbO\lambda}Q)^2 \langle Hb\rangle^2 x^2}{4V_{p\lambda}}\bigg|_{x=0}^{x=\hat{x}}, \quad (3.8)$$

where $Q = \frac{\langle HbO_2\rangle}{\langle Hb\rangle}$. Taking square root on both sides of equation (3.8), we get

$$\sqrt{v_{0\lambda}} = \frac{(\varepsilon_{Hb\lambda} + \varepsilon_{HbO\lambda}Q)\langle Hb\rangle x}{2\sqrt{V_{p\lambda}}}\bigg|_{x=0}^{x=\hat{x}} \quad (3.9)$$

It is easily seen that $\langle Hb\rangle x$ in equation (3.9) is the time varying quantity. During the period where $\langle Hb\rangle x$ varies linearly with time, equation (3.9) can be expressed as

$$\sqrt{\tilde{v}_{0\lambda}}\bigg|_{linear\ portion} = m_\lambda \langle Hb \rangle x \bigg|_{x=x_1}^{x=x_2}$$

Here m_λ is the slope of the linear portion and is given as $m_\lambda = \dfrac{(\varepsilon_{Hb\lambda} + \varepsilon_{HbO\lambda}Q)}{2\sqrt{V_{pR}}}$

Fig. 3.3 illustrates this slope concept. As an example calculation, one PPG cycle is processed as per the procedure discussed with a linear portion marked on it. Linear regression on the PPG samples of this linear portion can be carried out to estimate the best linear fit, from which the slope m_λ can be extracted. The slopes m_R and m_{IR} of the linear portions of the processed (as given in equation (3.9)) red and IR PPG signals are

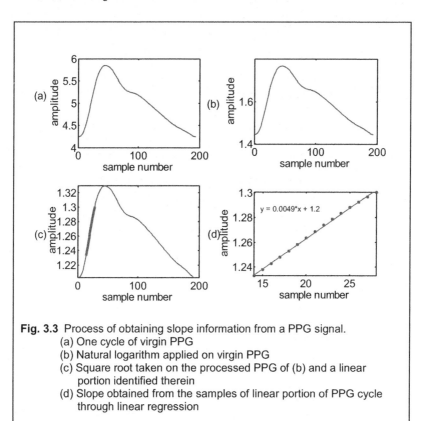

Fig. 3.3 Process of obtaining slope information from a PPG signal.
  (a) One cycle of virgin PPG
  (b) Natural logarithm applied on virgin PPG
  (c) Square root taken on the processed PPG of (b) and a linear portion identified therein
  (d) Slope obtained from the samples of linear portion of PPG cycle through linear regression

$$m_R = \frac{(\varepsilon_{HbR} + \varepsilon_{HbOR}Q)}{2\sqrt{V_{pR}}} \quad \text{and} \quad m_{IR} = \frac{(\varepsilon_{HbIR} + \varepsilon_{HbOIR}Q)}{2\sqrt{V_{pIR}}} \qquad (3.10)$$

The value of Q obtained from equation (3.10) is

$$Q = \frac{m_{IR}\sqrt{V_{pIR}}\varepsilon_{HbR} - m_R\sqrt{V_{pR}}\varepsilon_{HbIR}}{m_R\sqrt{V_{pR}}\varepsilon_{HbOIR} - m_{IR}\sqrt{V_{pIR}}\varepsilon_{HbOR}} \qquad (3.11)$$

Utilizing equations (3.11) and (1.3) we get

$$SpO_2 = \frac{m_{IR}\sqrt{V_{pIR}}\varepsilon_{HbR} - m_R\sqrt{V_{pR}}\varepsilon_{HbIR}}{\left[m_{IR}\sqrt{V_{pIR}}\varepsilon_{HbR} + m_R\sqrt{V_{pR}}\varepsilon_{HbOIR}\right] - \left[m_R\sqrt{V_{pR}}\varepsilon_{HbIR} + m_{IR}\sqrt{V_{pIR}}\varepsilon_{HbOR}\right]} 100\% \quad (3.12)$$

It is easily seen from equation (3.12) that the expression for computation of SpO_2 is independent of patient dependent variables as well as the intensities and sensitivities of the sources and detectors respectively. The details of experiments conducted to practically validate equation (3.12) are given next.

3.3 Experimental Results

The expression for SpO_2 is derived basically from equation (3.6) and hence it is necessary that the validity of that equation is tested by suitable experimentation. It is seen that equation (3.6) is independent of

(i) The intensities of the red and IR sources.

(ii) Sensitivity of the detector and the gain introduced by the signal conditioning electronics.

(iii) Patient dependent parameters.

To validate these three criteria, a clip-on sensor housing a red LED and an IR LED on one side and a photo diode detector on the other side was developed and interfaced to the signal conditioning circuitry shown in Fig. 3.4. The drives to the LEDs were time sliced at a rate of 1 kHz and the output from the photo diode is suitably de-multiplexed to separate the red and IR PPG signals as explained

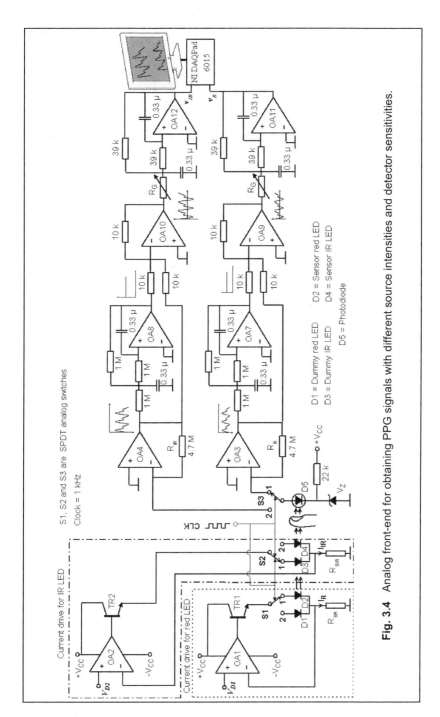

Fig. 3.4 Analog front-end for obtaining PPG signals with different source intensities and detector sensitivities.

in chapter 2. The red and IR PPG signals thus obtained are sampled at a rate of 200 sa/s and acquired using a 16-bit data acquisition card NI DAQPad-6015, manufactured by National Instruments [67]. The picture portrayed in Fig. 3.5 shows the experimental set up. Using the experimental setup, red and IR PPG signals were acquired from volunteers (after obtaining informed consent as per the directions of IIT Madras ethics committee). The volunteers participated in this study are healthy and non-smoking (mean age 30.6 years, SD 4.2 years, range 26-37 years). Data acquisition and processing were achieved under the LabVIEW program environment. The program developed is listed in Annexure 2. The acquired PPG signals are first filtered to remove high frequency noise.

3.3.1 Experimentation to Illustrate the Efficacy of the Proposed Method in Removing the Influence of Source Intensity

In order to demonstrate that applying natural logarithm to a PPG removes the influence of the source intensity, the currents ($I_R = V_{D1}/R_{SR}$ and $I_{IR} = V_{D2}/R_{SIR}$ in Fig. 3.4) through the LEDs (Red and IR) were varied over the range 2 mA and 8 mA in steps of 0.2 mA to obtain a variation in the optical output power in the range of 0.4 mW to 1.6 mW (optical intensity variation in the range of 3 mcd and 12 mcd). The red and IR PPG signals were recorded for each intensity value utilizing the experimental set-up shown in Fig.3.5. The experiment was performed on several volunteers in resting position. Thus, a total of 62 records of red and IR PPG signals at 31 different intensity levels were recorded per volunteer. In each case the duration of recording was set as 90 s. During the recording, the SpO_2 values were simultaneously obtained using a CPO, model planet 50, manufactured by Larsen and Toubro, India. To avoid interference, the sensor of CPO was positioned away (on a finger of the left hand of the volunteer) from the sensor (which was placed on a finger in the right hand) of the experimental setup.

Fig. 3.5 Complete experimental setup showing developed prototype PPG unit and data acquisition facility.

A snap shot taken during the recording of PPG signals from a volunteer is shown in Fig. 3.6. Natural logarithm of each PPG stream is then computed and the peak to peak amplitude values and slopes ascertained therefrom are indicated in Table 3.1. Table 3.1 lists measurements for seven intensity levels for brevity. It is clearly seen that even though the intensity is increased over fourfold, the peak-to-peak amplitude values and slopes of the processed PPGs, both red and IR, have remained constant. The small variations in the peak-to-peak values are due to the actual perfusion in the peripheral vessels (the amount of hemoglobin) changing in a random manner. Hence it is seen that equation (3.6) is indeed independent of the intensity of the source and patient dependent parameters. Fig. 3.7 illustrates that taking natural logarithm indeed eliminates the influence of source intensity. Traces (a) to (d) in Fig. 3.7 are IR PPG signals obtained from a volunteer at different source intensity levels. Trace (e) portrays all the signals in (a) through (d) after taking natural logarithm on each one of them. Fig. 3.7 clearly indicates that applying natural logarithm to a PPG signal indeed removes the influence of source intensity.

3.3.2 Experimentation to Illustrate the Efficacy of the Proposed Method in Removing the Influence of Detector Sensitivity

To demonstrate that applying natural logarithm to a PPG signal makes the PPG to become independent of detector sensitivity as well, PPG signals are recoded for various detector sensitivity levels. The intensity of the LED is set at 9 mcd (6 mA LED current, corresponding to 1.2 mW optical output power) and PPGs obtained with ten different detector sensitivity (gain) settings. The gain fixing resistor (R_G) in the last stage of the analog front end shown in Fig. 3.4 is suitably altered to achieve gains of 1, 5, 10, 20, 30, 40 and 50. The red and IR PPG signals at each of the gains were acquired and processed. It is evident from

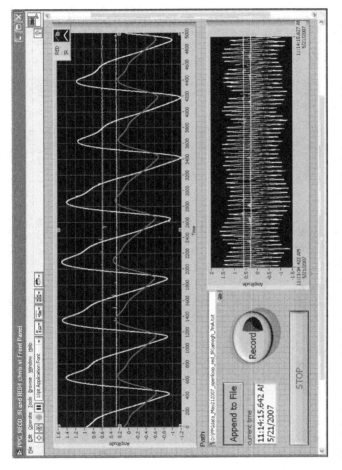

Fig. 3.6 Snapshot of the front panel of the prototype PPG, data recorded from a volunteer.

Table 3.1 Effect of processing on source intensity of the PPG signals obtained from a volunteer.

Source Intensity (mW)	Peak-to-peak amplitudes (V) before processing		Peak-to-peak amplitudes (V) after applying "ln"		Slopes before processing		Slopes after applying "ln"	
	V_{pR}	V_{pIR}	V_{pR}	V_{pIR}	m_R	m_{IR}	m_R	m_{IR}
0.4	0.06619	0.21313	0.09486	0.30585	0.00379	0.01178	0.00538	0.01729
0.6	0.11623	0.31672	0.10014	0.30962	0.00564	0.16216	0.00550	0.01716
0.8	0.16751	0.53433	0.09715	0.29669	0.00901	0.02778	0.00543	0.01749
1.0	0.22416	0.64992	0.09358	0.29468	0.01268	0.03918	0.00548	0.01751
1.2	0.27848	0.89114	0.09818	0.29735	0.01653	0.04577	0.00553	0.01788
1.4	0.31867	0.98862	0.09845	0.30466	0.01966	0.06172	0.00546	0.01781
1.6	0.36052	1.14654	0.10185	0.30335	0.02163	0.06459	0.00549	0.01768
Mean	*0.21882*	*0.6772*	*0.09774*	*0.30174*	*0.0127*	*0.05899*	*0.00546*	*0.01754*
Std. Dev	*0.01808*	*0.34869*	*0.00286*	*0.00555*	*0.0069*	*0.04908*	*0.00004*	*0.00026*

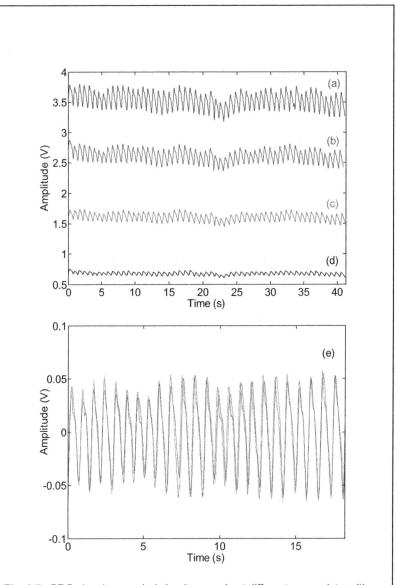

Fig. 3.7 PPG signals recorded simultaneously at different source intensities **(a)** 1.6 mW **(b)** 1.2 mW **(c)** 0.8 mW and **(d)** 0.4 mW
(e) Overlapped PPG signals after applying natural logarithm to traces (a) to (d)

Table 3.2 Effect of detector sensitivity on the processed PPG signals

Detector Gain	Peak-to-peak values (V)		Slopes	
	V_{pR}	V_{pIR}	m_R	m_{IR}
10	0.09818	0.29735	0.00553	0.01788
20	0.09818	0.29735	0.00553	0.01788
30	0.09818	0.29735	0.00553	0.01788
40	0.09818	0.29735	0.00553	0.01788
50	0.09818	0.29735	0.00553	0.01788
Mean	**0.09818**	**0.29735**	**0.00553**	**0.01788**
Std. Dev	**0.00000**	**0.00000**	**0.00000**	**0.00000**

results tabulated in Table 3.2 and the traces given in Fig. 3.8, that applying natural logarithm to a PPG signal removes the effect of detector sensitivity as well. It should be noted here that since breathing alters the PPG signal, the peak-to-peak amplitude of PPG cycles within a breathing cycle is averaged to remove the effect of breathing.

After establishing the facts that the processed PPG signals were independent of intensity of light source and detector sensitivity, the SpO_2 values for all the data obtained from different volunteers are estimated employing equation (3.12). Fig. 3.9 shows eighteen processed cycles of PPG data with the linear portions identified. Linear portions are selected from the systole portion of the processed PPG cycles. Linear regression analysis is carried out to determine the slope of selected linear portion of each of the cycles of Fig. 3.9, as depicted in the traces

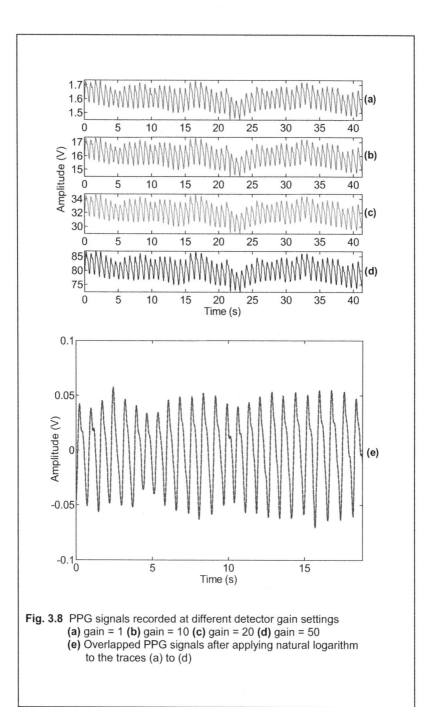

Fig. 3.8 PPG signals recorded at different detector gain settings
(a) gain = 1 (b) gain = 10 (c) gain = 20 (d) gain = 50
(e) Overlapped PPG signals after applying natural logarithm to the traces (a) to (d)

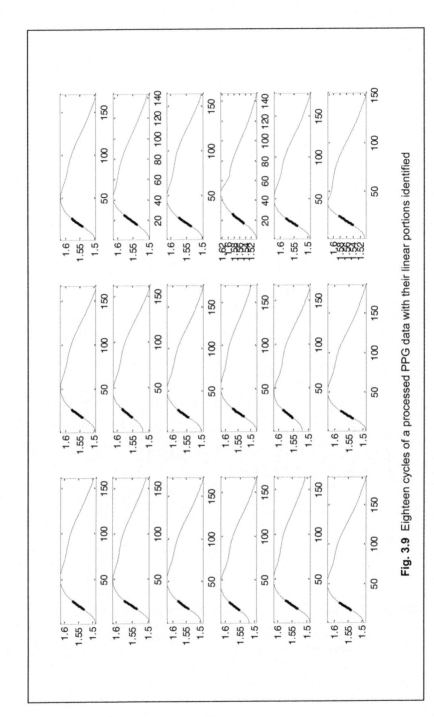

Fig. 3.9 Eighteen cycles of a processed PPG data with their linear portions identified

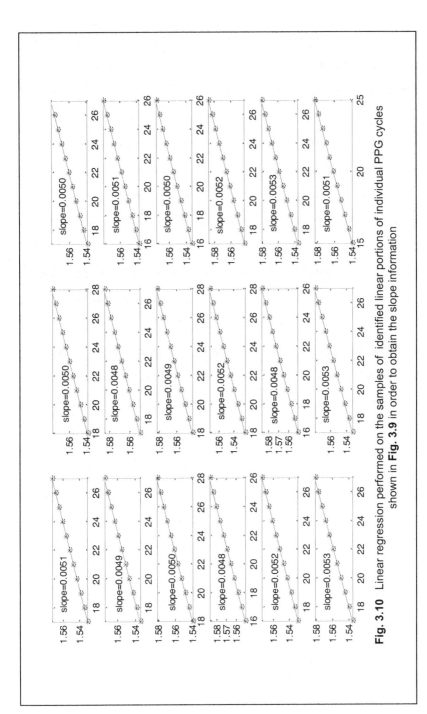

Fig. 3.10 Linear regression performed on the samples of identified linear portions of individual PPG cycles shown in **Fig. 3.9** in order to obtain the slope information

of Fig. 3.10. SpO_2 is then estimated using this slopes data and peak-to-peak amplitudes of the suitably processed PPG signals using equation (3.12). Fig. 3.11 portrays the snapshot of the front panel of the developed virtual instrument to accomplish the computation of SpO_2 utilizing equation (3.12). The SpO_2 was estimated on a cycle by cycle basis. A running mean over five consecutive values is taken and the resultant (moving window average) value of SpO_2 is updated every second on the front panel of the virtual instrument. The average SpO_2 values and the standard deviations (SD) of the data spread obtained using the proposed method are compared with the ones measured using a CPO in Table 3.3. It is seen that the proposed method provides stable readings which are comparable with the CPO even though the input intensity is varied over fourfold. The results presented in Table 3.3 also indicate that the computation of SpO_2, employing the proposed slope based method, is unaffected either by the sensor dependent parameters or patient dependent parameters. Hence reliance on a calibration curve, as is normally done for the pulse oximeters in vogue, is not essential.

Pulse oximeters are extensively calibrated to remove the effect of patient and sensor dependent parameters to obtain reliable SpO_2 readings. The method proposed in this chapter removes the effect of patient and sensor dependent parameters by appropriately processing the red and IR PPG signals. Required analytical expression is derived to compute the SpO_2 from the processed red and IR PPG signals. Experimental validation has been provided to establish that the proposed method indeed removes the influence of the source intensity and detector sensitivity. A prototype pulse oximeter was built under the LabVIEW environment and tested. The test results indicate that the proposed slope based method of computation of SpO_2 does not require the use of calibration curves.

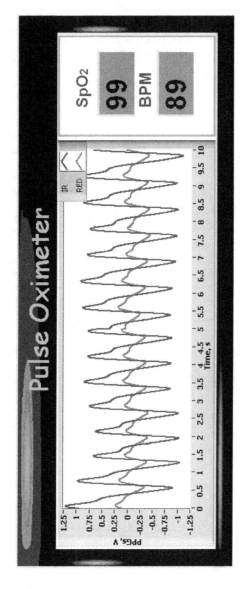

Fig. 3.11 Snapshot of the front panel of the prototype pulse oximeter employing the proposed slope based method.

Table 3.3 SpO$_2$ computed using the slope based method

Subjects	Average SpO$_2$ (Standard Deviation)			
	Proposed method with source intensity (mW)			CPO
	0.4	0.8	1.6	
1	98.16 (0.68)	97.98 (0.74)	98.36 (0.52)	**98.87 (0.34)**
2	98.00 (0.57)	98.31 (0.47)	98.70 (0.46)	**98.70 (0.44)**
3	98.87 (0.34)	99.14 (0.59)	98.69 (0.47)	**98.61 (0.24)**
4	98.38 (0.72)	97.81 (0.53)	97.98 (0.51)	**98.47 (0.31)**
5	97.88 (0.49)	98.26 (0.53)	98.62 (0.56)	**97.49 (0.50)**

It would be a welcome step if the steps in the processing are reduced resulting in a simpler method of pulse oximetry. Such a scheme is explained in the next chapter.

4. A Novel Peak Value Based Method of Measurement of SpO$_2$

4.1 Introduction

In the previous chapter, a method for the computation of SpO$_2$ from suitably processed PPG signals has been presented. SpO$_2$ values are computed utilizing the peak-to-peak amplitudes and slopes of the red and IR PPG signals after applying natural logarithm and square rooting operations. Though equation (3.12) of the slope based technique meets the set objective of obtaining a method of computation of SpO$_2$ to be independent of interfering parameters, the method is complex and laborious. First, the natural logarithm of the PPG is to be computed followed by square root operation. The peak-to-peak value of each PPG cycle is to be determined next. Linear portions of the processed PPG cycles are to be identified and the slopes of linear portions determined. Computation of SpO$_2$ is then performed using the slopes, peak-to-peak values and extinction coefficients. Hence, an alternate method meeting the objective yet simpler than the slope based method would be welcome. The peak value method described herein achieves just that.

4.2 The Peak Value Method of Estimation of SpO$_2$

The pulsatile component of the PPG signal after applying natural logarithm to a PPG obtained from equation (3.6) is

$$v_{0\lambda} = \ln(v_{0AC\lambda})\big|_{pulse} = -\frac{(\varepsilon_{Hb\lambda}\langle Hb \rangle + \varepsilon_{HbO\lambda}\langle HbO_2 \rangle)x^2}{2\hat{x}}\bigg|_{x=0}^{x=\hat{x}} \quad (4.1)$$

Applying equation (4.1) for the outputs of the red and IR photo detectors, we get the pulsatile portions v_{0R} and v_{0IR} as

$$v_{0R} = -\frac{(\varepsilon_{HbR}\langle Hb\rangle + \varepsilon_{HbOR}\langle HbO_2\rangle)x^2}{2\hat{x}}\bigg|_{x=0}^{x=\hat{x}}$$

$$v_{0IR} = -\frac{(\varepsilon_{HbIR}\langle Hb\rangle + \varepsilon_{HbOIR}\langle HbO_2\rangle)x^2}{2\hat{x}}\bigg|_{x=0}^{x=\hat{x}}$$

The peak-to-peak values V_{pR} and V_{pIR} of v_{0R} and v_{0IR} respectively are:

$$V_{pR} = \frac{(\varepsilon_{HbR}\langle Hb\rangle + \varepsilon_{HbOR}\langle HbO_2\rangle)\hat{x}}{2} \qquad (4.2)$$

$$V_{pIR} = \frac{(\varepsilon_{HbIR}\langle Hb\rangle + \varepsilon_{HbOIR}\langle HbO_2\rangle)\hat{x}}{2} \qquad (4.3)$$

Dividing equation (4.2) by (4.3) results in

$$\frac{V_{pR}}{V_{pIR}} = \frac{(\varepsilon_{HbR}\langle Hb\rangle + \varepsilon_{HbOR}\langle HbO_2\rangle)}{(\varepsilon_{HbIR}\langle Hb\rangle + \varepsilon_{HbOIR}\langle HbO_2\rangle)} = \frac{(\varepsilon_{HbR} + \varepsilon_{HbOR}Q)}{(\varepsilon_{HbIR} + \varepsilon_{HbOIR}Q)},$$

where $Q = \dfrac{\langle HbO_2\rangle}{\langle Hb\rangle}$. Rearranging the above equation, we get

$$Q = \left(\frac{V_{pIR}\,\varepsilon_{HbR} - V_{pR}\,\varepsilon_{HbIR}}{V_{pR}\,\varepsilon_{HbOIR} - V_{pIR}\,\varepsilon_{HbOR}}\right) \qquad (4.4)$$

Substituting Q from equation (4.4) in equation (1.3) we get

$$SpO_2 = \left(\frac{V_{pIR}\,\varepsilon_{HbR} - V_{pR}\,\varepsilon_{HbIR}}{(V_{pR}\,\varepsilon_{HbOIR} - V_{pIR}\,\varepsilon_{HbOR}) + (V_{pIR}\,\varepsilon_{HbR} - V_{pR}\,\varepsilon_{HbIR})}\right)100\ \% \qquad (4.5)$$

It is easily seen that equation (4.5) is devoid of not only patient dependent parameters but also independent of the red and IR source intensities and detector sensitivity. Since equation (4.5) stems from the basic equation (3.6), the experiments conducted as per sections 3.3.1 and 3.3.2 are also valid in establishing equation (4.5). Hence equation (4.5) indeed is free of source intensities and detector sensitivities.

4.2.1 Error in the Proposed Method of Measurement of SpO$_2$

The inherent error δ_{SpO2} in computing SpO$_2$ as per equation (4.5) in terms of errors δ_{VR} and δ_{VIR} in the measurement of V_{pR} and V_{pIR} and errors δ_1, δ_2, δ_3, and δ_4 in ε_{HbR}, ε_{HbOR}, ε_{HbIR}, and ε_{HbOIR} is derived as

$$\delta_{SpO2} = \frac{p}{q-p}(\delta p - \delta q) \qquad (4.6)$$

Where $p = V_{pR}\varepsilon_{HbOIR} - V_{pIR}\varepsilon_{HbOR}$, $q = V_{pR}\varepsilon_{HbIR} - V_{pIR}\varepsilon_{HbR}$

$$\delta p = \left(V_{pR}\varepsilon_{HbOIR}\left(\delta_{VR} + \delta_4\right) - V_{pIR}\varepsilon_{HbOR}\left(\delta_{VIR} + \delta_2\right)\right)/p$$

$$\delta q = \left(V_{pR}\varepsilon_{HbIR}\left(\delta_{VR} + \delta_3\right) - V_{pIR}\varepsilon_{HbR}\left(\delta_{VIR} + \delta_1\right)\right)/q$$

Here, δ_{VR} and δ_{VIR} can be made insignificant by employing a high resolution ADC. For the prototype, a 16-bit ADC with a measurement error of 0.01 % is employed. It is found that δ_1, δ_2, δ_3, and δ_4 are the main parameters influencing the error in estimation of SpO$_2$ due to mismatch between the extinction coefficients at rated wavelengths and actual wavelengths emitted by red and IR LEDs. The worst case error due to this is estimated to be less than 1%.

4.3 Experimental Results

To ascertain the feasibility of the proposed scheme, the PPG data acquired from the experiments conducted on volunteers(mean age 30.6 years, SD 4.2 years, range 26-37), as mentioned in the previous chapter, are used here for the estimation of SpO$_2$ utilizing equation (4.5). Table 4.1 lists the results of the estimation obtained from volunteers' data at different LED intensities reiterating that equation (4.5) is devoid of sensor and patient dependent parameters. They

Table 4.1 SpO$_2$ computed using peak value based method

Subjects	Average SpO$_2$ (Standard Deviation)			
	Proposed method with source intensity (mW)			CPO
	0.4	0.8	1.6	
1	97.52 (0.36)	97.70 (0.55)	98.05 (0.39)	**98.87 (0.34)**
2	98.57 (0.45)	97.93 (0.34)	97.88 (0.24)	**98.70 (0.44)**
3	98.02 (0.30)	98.12 (0.34)	97.96 (0.38)	**98.61 (0.24)**
4	97.56 (0.50)	98.05 (0.36)	97.79 (0.42)	**98.47 (0.31)**
5	97.31 (0.68)	97.83 (0.41)	98.11 (0.38)	**97.49 (0.50)**

are in good agreement with the SpO$_2$ values obtained from the CPO. Fig. 4.1 portrays the snapshot of the front panel of the developed virtual pulse oximeter instrument to accomplish the estimation of SpO$_2$ utilizing equation (4.5). Fig. 4.2 shows the SpO$_2$ estimation using the proposed method and traditional method (equation (1.9)) when applied to the data obtained from a particular volunteer. The readings of CPO, which were noted down every second at the time of PPG signal recording from the volunteer, are also displayed on the same plot for comparison. It can be seen that the estimations utilizing the proposed peak value based method closely match with the ones obtained from the CPO. It is seen that equation (4.5) possesses all the advantages of equation (3.12) of slope based method while reducing the number of operations to be carried out to process the

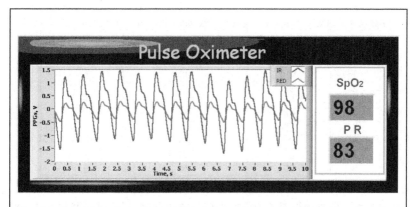

Fig. 4.1 Snapshot of the front panel of the prototype pulse oximeter based on the proposed peak value method

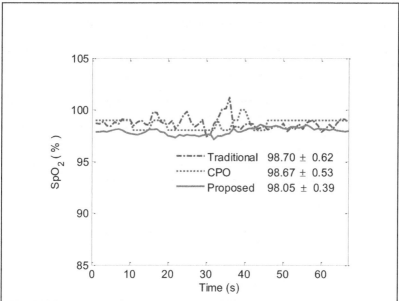

Fig. 4.2 Estimated SpO$_2$ obtained from traditional method (blue, dash-dotted line), readings of commercial pulse oximeter (red, dotted line) and proposed peak value based method (violet, solid line).

Table 4.2 SpO$_2$ computed using peak value based method on L&T data

S. No	SpO$_2$ of L&T Data	SpO$_2$ estimated from proposed method	Error (%)
1	100	98	-2.00
2	98	96	-2.04
3	96	94	-2.08
4	94	93	-1.06
5	92	92	-1.08
6	90	90	0.00
7	85	86	+1.18
8	80	84	+5.00

red and IR PPG signals. Hence the peak value method is better suited for on-line measurement of SpO$_2$.

Here too the red and IR PPG data for SpO$_2$ ranging from 80 to 100 % obtained from L&T Medical Systems, Mysore, India, was used to test the efficacy of the proposed method. The estimated SpO$_2$ values obtained employing the proposed peak value method, rounded-off to nearest integer are listed in Table 4.2. Preliminary results on applicability of proposed method for below normal saturation values indicate that the error is within acceptable limits up to SpO$_2$ of 85%. But, for SpO$_2$ below 85%, the error seems to be higher than that claimed by available CPOs. However, further investigations need to be done under clinical settings on volunteers for saturations ranging from 70 to 100%.

All the three proposed methods of SpO$_2$ estimation, as detailed in chapter 2, 3, and 4 namely model based method, slope based method and peak value based method respectively, are applied on red and IR PPG signals. Test results on two subjects are presented Annexure 3.

In chapter 1, it was pointed out that the reliability of pulse oximeters is very much affected by motion induced artifacts. A novel solution to reduce motion artifacts from corrupted PPG signals is portrayed in the next chapter.

5. Motion Artifact Reduction in PPG Signals using Singular Value Decomposition

As brought out in earlier chapters, in pulse oximetry, computation of SpO_2 requires red and IR PPG signals. A PPG signal obtained using a sensor (either of the transmission type or of the receiver type) is dependant on the light coupled from the source to the body as well as the light coupled from the body to the detector. To achieve proper coupling, the source and the detector should be in close contact with the body. To ensure close contact, the sensor housing the sources and detectors should apply sufficient pressure. Any pressure applied by the sensor increases the temperature of the region underneath the sensor leading to sweating and discomfort to the patient. Moreover, the sweating and increased temperature would severely alter the blood perfusion, resulting in erroneous PPG signals. Hence the sensor housing the red and IR sources and detectors in any pulse oximeter is designed to exert just bare minimum pressure required to make a contact between the body and the sensor. If a patient connected to a pulse oximeter moves, the contact between the sensor and the patient's body gets affected, corrupting the PPG signals obtained during such periods of movement with motion artifacts. SpO_2 computed using the motion artifact corrupted PPG signals would be erroneous [46], [47], [68]. Most pulse oximeters manufacturers solve this problem by simply suppressing the output (with or without indication on the front panel of the oximeter) during such periods. Some pulse oximeters simply display the last valid SpO_2 reading available during such periods of movements [72]. Even this simple technique of suppression during periods, where the computation of SpO_2 is not possible, requires that the oximeters are made capable of recognizing such periods. In other words, most of the processing in present day pulse oximeters is geared towards artifact recognition rather than artifact reduction. Hence, processing of PPG signals with a view to remove or reduce motion artifacts, if any, present in the PPG signals is of particular concern within the context of pulse oximeters.

There have been many attempts to reduce the influence of motion artifacts from corrupted PPG signals. The most common technique employed for the reduction of the motion artifact is the moving average method [9], [38]. The moving average method works well only for a limited range of artifacts. In-band noise results when the spectra of motion artifact and that of the PPG signal overlap significantly. It has been shown that the in-band noise can be successfully reduced with the use of adaptive filters [69]-[72]. However, to apply adaptive filter technique, a reference signal that is strongly correlated either with the artifact but uncorrelated with the signal or strongly correlated with the signal but uncorrelated with artifact must be available. In most cases, suitable reference signals, representing motion artifact were obtained by employing additional hardware. For example, reference signals were obtained from an additional transducer attached to sense the movement [69], [70] or by employing an additional reflectance type optoelectronic sensor [71]. Use of a synthetic reference signal estimated from the artifact-free part of the PPG signal [72] was also reported for reducing motion artifact. The Masimo SET® [73] identified venous blood volume change as a significant contributor to noise during motion and hence in their method a venous noise reference signal is extracted from the artifact-induced PPG signal itself, without extra hardware. The added artifact is then removed from the PPG signal using adaptive noise cancellation. A signal processing technique was also suggested using multi-rate filter bank and a matched filter [74] that has better performance compared to the moving average and adaptive filtering approaches. The dynamic nature of the biological systems causes most biological signals to be non-stationary and change substantially in their properties over time. Making use of this non-stationary nature of the PPG signals, time-frequency methods like wavelet transforms [75] and smoothed pseudo Wigner-Ville distribution [76] have been applied on PPG signals to provide significant improvements compared to traditional approaches. A model-based artifact reduction methodology [77] was proposed with a nonlinear optical receiver based upon inversion of a physical artifact model. This approach has been implemented [78] utilizing an additional source-detector pair, resulting in the three wavelength probe, for reduction of motion artifacts. It has also been demonstrated that independent component analysis (ICA) can be applied to reduce the motion artifacts by exploiting the independence between PPG and motion artifact signals. While the third order ICA of the time-derivative of PPG signals resulted in better artifact suppression for pulse oximetry [79], the ICA applied with a pre-processing called block interleaving with low pass filtering performed better than ICA alone [80]. But a study directed on the statistical independence of motion artifacts, arterial and venous components of a PPG signal [81] indicated that the arterial pulsations are not statistically independent from motion. A very recent comparison study on the efficacy of wavelet transform and adaptive filtering techniques in restoring the artifact induced PPG signals [82] for estimation of heart rate (HR) and pulse transit time (PTT) revealed that both methods introduce phase shifts to the PPG signals and concluded that the wavelet transform technique has limited application in restoring motion artifact corrupted PPG signals. Hence, further research aimed at improving the performance of motion artifact rejection is still required.

In this chapter, an artifact reduction method, applicable for PPG signals, based on singular value decomposition (SVD), is presented. The proposed method extracts clean artifact-free PPG signals from artifact riddled PPG signals preserving all the essential morphological features required [83].

5.1 SVD for Motion Artifact Reduction

5.1.1 Singular Value Decomposition

Singular value decomposition, proposed in 1870, is an important tool of linear algebra. Let X be an $m \times n$ matrix of real valued data, with $m \geq n$ and hence rank (p) of $X \leq n$. Then singular value decomposition of matrix X is given by [84]

$$X = USV^T \qquad , \qquad (5.1)$$

where U is an $m \times n$ matrix, S is an $n \times n$ diagonal matrix ($S = diag(\sigma_1,...,\sigma_n)$) and V^T is also an $n \times n$ matrix. The diagonal elements of S are called singular values. The singular values are the positive square roots of the eigen values of X^TX. Also $\sigma_q > 0$ for $1 \leq q \leq p$ and $\sigma_q = 0$ for $(p+1) \leq q \leq n$. U and V are unitary matrices with $U^TU = I$ and $V^TV = I$, where I is an identity matrix. The columns of U are called left singular vectors of X, while the columns of V are called right singular vectors of X. An important observation is that as the singular values decay rapidly, with $\sigma_1 \geq \sigma_2 \geq ... \geq \sigma_n \geq 0$, we can expect that there will be a good lower rank approximation ($\hat{X}$) to X by setting the small singular values to zero. The lower rank approximation $\hat{X}$ is given by

$$\hat{X} = \sum_{i=1}^{j} u_i \sigma_i v_i^T \qquad (5.2)$$

where σ_i assumed to be zero for $i > j$; u_i and v_i are the jth columns of U and V respectively. The fact that singular values of a given data matrix contains information about the noise level in the data, energy and rank of the matrix are exploited for signal processing (data compression, noise removal and pattern extraction) [85]. This feature is exploited here to remove motion artifacts from corrupted PPG signals.

5.1.2 Principal Component Extraction using SVD

It can be easily shown that if samples of a perfectly periodic waveform $x(k)$, $k=1,2,...,mn$ with a period length n samples is formed into a data matrix X by placing data pertaining to each period of length n as a row of X as given below,

$$X = \begin{bmatrix} x(1) & x(2) & ... & x(n) \\ x(n+1) & x(n+2) & ... & x(2n) \\ \vdots & \vdots & & \vdots \\ x((m-1)n+1) & x((m-1)n+2) & ... & x(mn) \end{bmatrix} \quad (5.3)$$

then the SVD of X will be a rank one matrix, where σ_1 only be nonzero and σ_2 to σ_n will all be zero. Thus a dominant first singular value obtained from the SVD of a given data matrix indicates a strong periodic component in the rows of the data matrix. In such a case, the periodic signal is enshrined in $\sigma_1 u_1 v_1^T$, where u_1 and v_1 are the first columns of the corresponding left and right singular vectors.

However, for a quasi-periodic signal like PPG, σ_2 to σ_n would never become zero. On the other hand, if a particular matrix X_r with a row length r that matches with the dominant frequency of the given PPG, then the ratio of first two singular values σ_{1r}/σ_{2r} for X_r will be a maximum. This fact is the basis for the method being presented here, for extracting an artifact-free PPG from a corrupted PPG.

The procedure is:

Let $x(k)$, $k = 1...M$ represent the sampled values of the PPG signal to be processed. Matrices $X_1, X_2...X_r...$, of different row lengths are formed using the data $x(k)$. The range of row lengths are chosen to represent expected range of heart rate, say, between 0.8 Hz to 2 Hz. SVD is performed on each matrix and ratio of the first two singular values, σ_1/σ_2, called singular value ratio (SVR), is computed in each case. The ratios are then plotted against the row length to obtain a graph called the SVR spectrum of the signal. From the SVR spectrum, the particular value of row length, say r, for which the SVR (σ_{1r}/σ_{2r}) is maximum is obtained and the corresponding data matrix X_r is selected to represent the data. It is evident then that the row length, say r, represents the dominant periodicity of the PPG. From the SVD of X_r, one cycle representing the dominant PPG signal is reconstructed by taking the average of all rows of $\sigma_1 u_1 v_1^T$. It should be noted here that the averaging process eliminates noise and artifact present in the signal and extracts only the signal corresponding to the dominant periodicity represented by the row length r. Once one cycle of the signal is extracted from $x(k)$, $k = 1...M$ then the first r values $x(1)$ to $x(r)$ are dropped from $x(k)$ and r new values from the samples of the raw PPG appended to $x(k)$ at the end and the new data $x(k)$, $k = (r+1)...(M+r)$ is obtained. The whole procedure of obtaining an SVR spectrum is repeated with the new data set, and the second PPG cycle is extracted from that data matrix that exhibits maximum value of SVR. After extraction, once again the first row of data is dropped and an equal number of samples appended at the end to get a new data set. The entire process is repeated in a continuous manner and the processed PPG is obtained cycle after cycle incessantly from the samples of the raw PPG.

5.2 Experimental Results

In order to test the practicality of the SVD technique for motion artifact reduction, a prototype pulse oximeter was developed incorporating the analog front-end circuit developed in chapter 2. The analog red and IR PPG signals are sampled at a rate of 250 sa/s using National Instruments NIDAQPad-6015 data acquisition card. The proposed SVD algorithm is implemented in the Matlab environment. To demonstrate that the proposed method does not alter the morphological features of the PPG, first the SVD technique is applied on an artifact-free PPG. Data $_1x(k)$, k = 1, 2,..., 1250 for 5 seconds duration, which roughly includes 6 cycles of PPG signal as indicated in Fig. 5.1(a), is taken up for processing. First, the data is formed into a matrix, say $_1X_1$, with a row length of 300 samples depicting a heart rate of 0.833 Hz (50 beats per minute, bpm). SVD is then performed on this data matrix $_1X_1$ and the SVR, say SVR$_1$ computed therefrom. The data is then reformed into a second matrix $_1X_2$ with a row length of 299 depicting a heart rate of 0.836 Hz and the second SVR for this matrix, say SVR$_2$ is obtained. This process is continued with a decrease in row length (increasing effective heart rate) in steps of 1 sample up to a row length of 125 samples depicting a heart rate of 2 Hz (120 bpm). The SVR values obtained in each case are then plotted to obtain the first SVR spectrum with the range of periodicity from 0.833 Hz to 2 Hz (heart rate variations between 50 bpm to 120 bpm) as shown in Fig. 5.2. It is seen that the peak of this first SVR spectrum, shown as solid blue line, occurs at a row length of 186 samples (heart rate of 1.344 Hz). Hence the SVD of that matrix with its row length 186 depicting the heart rate of 1.344 Hz is taken and the first PPG cycle of the processed PPG shown in Fig. 5.1(b) is extracted from the $\sigma_1 u_1 v_1^T$ of that matrix. Once the first cycle is extracted, the first 186 samples are dropped from the original 1250 data samples and 186 new samples (from 5 s to 6 s of the raw data) are appended at the end to form a new data set $_2x(k)$ for further processing. The SVR spectrum for this data is determined exactly similar to the one obtained for $_1X_n$. The SVR

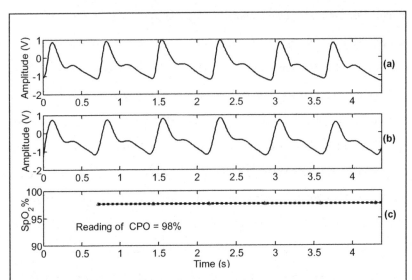

Fig.5.1 (a) Artifact-free red PPG as acquired (for simplicity only red PPG is shown) **(b)** Recovered red PPG using the proposed SVD technique **(c)** SpO_2 computed from the original PPG signals (dotted line) and recovered PPG signals (solid line)

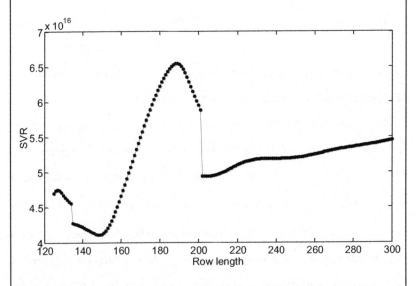

Fig. 5.2 SVR spectrum obtained during the reconstruction of first cycle of PPG signal shown in Fig. 5.1(a)

spectrum obtained for $_2X_n$ is portrayed in Fig. 5.3. The second cycle of the PPG is recovered by employing that matrix $_2X_r$, where the new row length r is 189 (corresponding to a heart rate of 1.323 Hz) samples as seen in Fig. 5.3, the row length for which the SVR is maximum. Likewise, the SVD processed PPG shown in Fig. 5.1(b) is recovered from the raw PPG cycle-by-cycle. Fig. 5.3 portrays all the six SVR spectra corresponding to the six recovered cycles of PPG shown in Fig. 5.1(b). It should be noted here that the yth cycle of the reconstructed PPG enshrines $(y+r-1)$ cycles of the raw PPG, r the row length of the matrix corresponding to maximum SVR. It should also be noted that when a matrix is reformed with a new row length, invariably some tail end data would be in excess. The excess data is simply discarded. For example, if a data length of 1250 samples is formed with a row length of 200 samples, after forming the sixth row, 50 samples would be left out, and are discarded. Visual comparison of the raw PPG of Fig. 5.1(a) and the processed PPG of Fig. 5.1(b) indicates that the method preserves the essential morphological features including the dicrotic notch. To analytically establish that the proposed SVD method does not introduce additional errors, two tests were carried out. First the normalized root mean square error (*NRMSE*) in the recovered PPG, x_R, is evaluated as

$$NRMSE = 20\log\left(\sqrt{\frac{\sum_{i=1}^{r}(x(i)-x_R(i))^2}{\sum_{i=1}^{r}(x(i))^2}}\right) \text{ dB} \qquad (5.4)$$

The *NRMSE* is found to be - 26.5 dB. Secondly, SpO$_2$ is estimated from the raw and processed PPG signals utilizing equation (4.5). The evaluated SpO$_2$ values are plotted in Fig. 5.1(c). It is seen that the SpO$_2$ values estimated from the raw and processed PPG signals match exactly thus illustrating the fact that the proposed SVD based method does not introduce additional errors when dealing

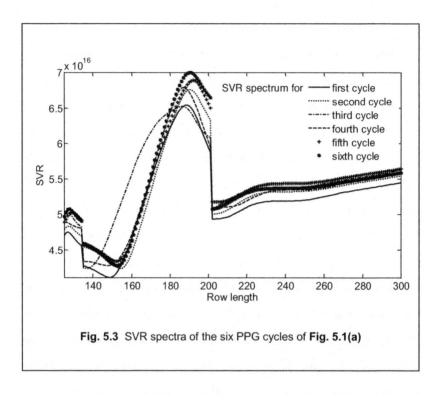

Fig. 5.3 SVR spectra of the six PPG cycles of **Fig. 5.1(a)**

with PPG signals uncorrupted with motion artifact. Fig. 5.1(c) also portrays the SpO_2 (single value) indicated by a CPO, model Planet 50, for comparison.

To evaluate the performance of the proposed method, frequently encountered artifacts during post-operative recovery and intensive care were created intentionally during the PPG data recording process. The experimental protocol followed for recording the data is:

- The sensor is connected to the left hand of the volunteer.
- Each recording of red and IR PPG is for a span of four minutes.
- Initial one minute of PPG is recorded without motion artifact by keeping the finger, connected to the sensor probe, at rest.

- At the end of the first minute, the volunteer is asked to intentionally introduce artifact by moving the finger in a particular direction, say, vertical for two minutes.
- Recording ended with a finial one minute period wherein the finger again is kept at rest without any motion.
- During the period of testing, the SpO_2 values indicated by a CPO are noted down by connecting the senor of the CPO to a finger on the right hand (kept motion free) of the volunteer. These SpO_2 values are used for comparing estimated SpO_2 after reducing the artifacts from corrupted PPG signals.

This protocol is repeated on each volunteer tested for each of the following four movements which are encountered in reality, namely, (i) vertical motion, (ii) horizontal motion, (iii) bending of the finger and (iv) applying pressure on the sensor (pressing the finger). Experiments were carried out on ten volunteers, after obtaining an "informed consent". The procedure was also approved by the ethics committee of IIT Madras. Forty such data sets were collected in total from the volunteers.

The algorithm was applied on all these artifact corrupted PPG signals. In every case, proper artifact-free PPG signals were recovered from the corrupted ones. Effectiveness of the proposed artifact reduction method is visible in Fig. 5.4 to Fig. 5.7 for artifacts induced due to horizontal motion, applying pressure, vertical motion and bending of finger respectively. Visual inspection of all the figures indicates that in all the cases artifacts are significantly reduced by the proposed method. Fig. 5.8 illustrates a sample PPG with a typical artifact, the recovered PPG using the proposed SVD based procedure and the SpO_2 values

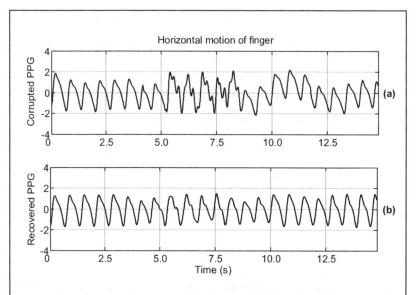

Fig. 5.4 (a) Artifact corrupted PPG due to horizontal motion of finger
(b) Recovered PPG using proposed SVD based method

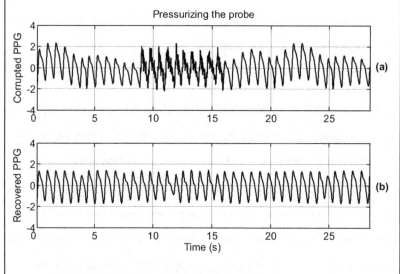

Fig. 5.5 (a) Artifact induced PPG by pressurizing the probe
(b) Recovered PPG using proposed SVD method

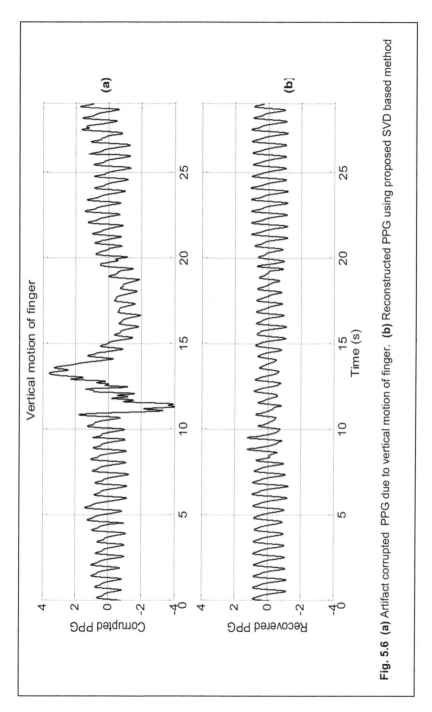

Fig. 5.6 (a) Artifact corrupted PPG due to vertical motion of finger. (b) Reconstructed PPG using proposed SVD based method

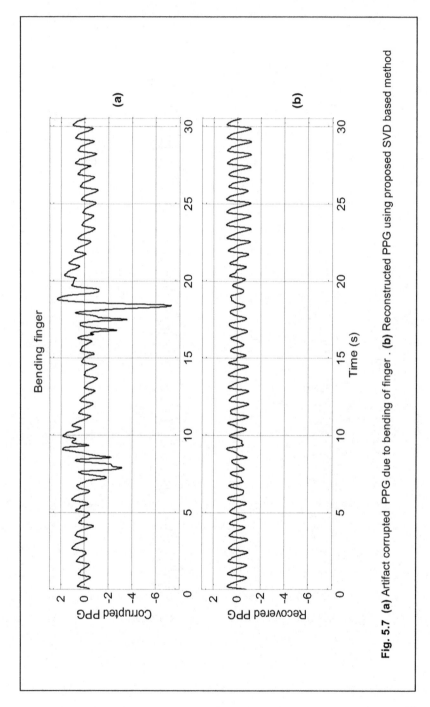

Fig. 5.7 (a) Artifact corrupted PPG due to bending of finger. (b) Reconstructed PPG using proposed SVD based method

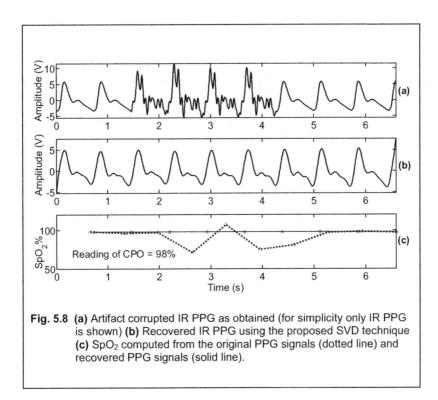

Fig. 5.8 (a) Artifact corrupted IR PPG as obtained (for simplicity only IR PPG is shown) **(b)** Recovered IR PPG using the proposed SVD technique **(c)** SpO_2 computed from the original PPG signals (dotted line) and recovered PPG signals (solid line).

computed before and after applying the proposed method. The SVR spectrum obtained in the process of applying the proposed algorithm on the corrupted PPG of Fig. 5.8(a) is given in Fig. 5.9. The efficacy of the algorithm can be clearly seen in the error-free estimation of SpO_2 even when the signals are contaminated with motion artifacts. During the controlled experiment with artifact depicted in Fig. 5.8, SpO_2 value obtained simultaneously with the help of CPO as explained earlier is also indicated in Fig. 5.8. It is seen that the SpO_2 values evaluated by the present method compares well within 2% of that indicated by the CPO. Hence it can be concluded that the filtering inherent in the SVD algorithm yields a significant improvement in the performance of motion artifact rejection.

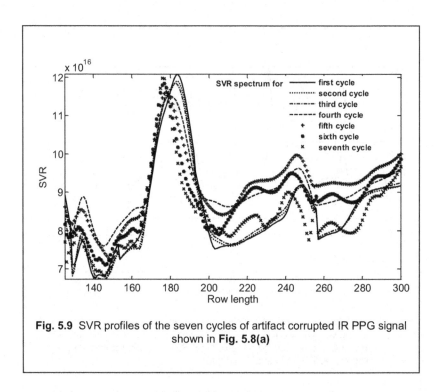

Fig. 5.9 SVR profiles of the seven cycles of artifact corrupted IR PPG signal shown in **Fig. 5.8(a)**

Though the SVD based technique works well in reducing motion artifacts from corrupted PPG signals, the method is computationally very intensive. An alternate method that is less computationally intensive compared to the SVD technique is developed in the next chapter.

6. Fourier Series Analysis for Motion Artifact Reduction and Data Compression of PPG Signals

6.1 Need for Alternative Method for Motion Artifact Reduction

As mentioned in chapter 5, the performance of pulse oximeters is affected by the movement of the patient connected to an oximeter. Even minor movements corrupt the PPG signals with severe artifacts and render the PPG signals during those periods unusable for estimation of oxygen saturation. As already pointed out a number of approaches [9], [38], [69]-[82] for the reduction of motion artifacts have been proposed, including a new method based on SVD [83] that is presented in chapter 5. However, all these methods suffer from one drawback or the other and hence the race is still on to find a most appropriate method that is economical in terms of complexity of operation and computation requirements. Such a simple method should also provide optimum reduction of motion artifacts recovering a clean PPG preserving all the morphological features from a corrupted PPG. This chapter discusses a computationally efficient method to reduce artifacts from PPG signal by applying the well known Fourier series analysis (FSA) on a cycle-by-cycle basis. It is seen that, as a result of the proposed analysis, each cycle of the PPG is represented by a reduced set of Fourier series coefficients resulting in data compression as well.

6.2 Cycle-by-cycle FSA Applied to PPG signals

It is well known that any periodic signal can be decomposed into a set of sinusoids made of a fundamental frequency and its harmonics as described by the Fourier series [86]. If $f(t)$ represents the periodic signal with a period T, then the Fourier series expansion of $f(t)$ is

$$f(t) = a_0 + \sum_{k=1}^{\infty} a_k \cos(k\omega t) + b_k \sin(k\omega t)$$

where $\omega = \dfrac{2\pi}{T}$ and

$$a_0 = \frac{1}{T} \int_0^T f(t)\, dt \tag{6.1}$$

$$a_k = \frac{2}{T} \int_0^T f(t) \cos(k\omega t)\, dt \tag{6.2}$$

$$b_k = \frac{2}{T} \int_0^T f(t) \sin(k\omega t)\, dt. \tag{6.3}$$

However, Fourier series is applicable only to periodic signals and hence can not be directly applied to a PPG signal which is quasi-periodic and non stationary. In the method being presented here, this problem is overcome by applying Fourier series on the PPG signal, on a cycle-by-cycle basis. First a complete cycle of a PPG as indicated in Fig. 6.1(a) is identified and its time period, say T_1, is determined. Assuming that this first cycle repeats itself endlessly, Fourier coefficients, say, $_1a_0$, $_1a_k$ and $_1b_k \big|_{k=1,2,3..\infty}$ are computed using equations (6.1), (6.2) and (6.3) with $T = T_1$. It should be noted here that the coefficients $_1a_0$, $_1a_k$ and $_1b_k \big|_{k=1,2,3..\infty}$ are strictly applicable only to the first cycle that is being processed, since in a typical PPG, the time periods of successive cycles will vary as set by the heart rate variability of the patient being tested. Once the Fourier coefficients applicable for the first PPG cycle are computed and stored, the second cycle in the PPG signal (with its time period T_2) is then identified and

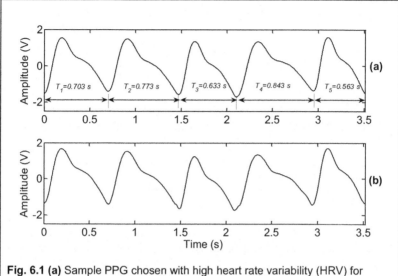

Fig. 6.1 (a) Sample PPG chosen with high heart rate variability (HRV) for testing the efficacy of the proposed method.

(b) The reconstructed PPG obtained after applying the proposed method on **(a)**. It is easily seen that the FSA method performs well even with high HRV and hence is insensitive to variations in heart rate.

subjected to Fourier series expansion and the Fourier coefficients $_2a_0$, $_2a_k$ and $_2b_k \mid_{k=1,2,3..\infty}$ again applicable only to that cycle are evaluated. This process is repeated for every cycle and in general, the mth cycle will be represented by its set of coefficients $_ma_0$, $_ma_k$ and $_mb_k \mid_{k=1,2,3..\infty}$.

A reverse process is applied to reconstruct the PPG signal $f_R(t)$ from the stored set of coefficients cycle-by-cycle. The first cycle $f_R(t)\mid_{t=0\ to\ T_1}$ is reconstructed as

$$f_R(t)\mid_{t=0\ to\ T_1} = {_1a_0} + \sum_{k=1}^{\infty} {_1a_k}\ cos\ (k\ \omega_1 t) + {_1b_k}\ sin\ (k\ \omega_1 t), \qquad (6.4)$$

where $\omega_1 = \dfrac{2\pi}{T_1}$. Similarly each cycle is then reconstructed one after the other and in general the mth cycle is reconstructed as

$$f_R(t)\big|_{t=T_{(m-1)} \text{ to } T_m} = {}_m a_0 + \sum_{k=1}^{\infty} {}_m a_k \cos(k\,\omega_m t) + {}_m b_k \sin(k\,\omega_m t), \qquad (6.5)$$

where $\omega_m = \dfrac{2\pi}{T_m}$. To demonstrate that the proposed method is insensitive to heart rate variation, the PPG shown in Fig. 6.1(a) is intentionally chosen such that successive cycles have widely differing time periods.

Even though the range of k in all the above expressions is from 1 to ∞, for a practical case, only a few significant terms need to be considered and hence the maximum value of k, say k_{max}, would be within a practical limit. The first ten significant Fourier coefficients of the five cycles of the sample PPG shown in Fig. 6.1(a) are listed in Table 6.1. From Table 6.1 it is seen that the magnitudes of the Fourier coefficients of higher order terms ($k > 7$) diminish drastically. However, restricting the range of k would introduce errors in the reconstructed signal and may change the most important aspect of the PPG, namely, the morphological features. If we restrict the number of Fourier coefficients to k_{max} in the reconstruction, then obviously the reconstructed signal will not be the same as the original signal and will deviate from the original signal. The deviation of the reconstructed PPG from the original PPG is quantified by the normalized root mean squared error ($NRMSE$) in the reconstructed signal. $NRMSE$ can be calculated as

$$NRMSE = 20\log\left(\sqrt{\dfrac{\int_0^{T_m}(f(t)-f_R(t))^2}{\int_0^{T_m}(f(t))^2}}\right) \text{ dB} \qquad (6.6)$$

Table 6.1 Fourier coefficients of sample PPG cycles of **Fig. 6.1(a)**

Coeff.	Cycle 1	Cycle 2	Cycle 3	Cycle 4	Cycle 5
a_0	0.25478	0.06097	-0.24276	0.01994	0.22472
a_1	-0.87716	-0.82329	-0.82483	-0.84233	-0.83908
a_2	-0.58802	-0.57765	-0.57629	-0.58082	-0.58030
a_3	-0.09580	-0.09116	-0.09047	-0.09281	-0.09226
a_4	-0.02790	-0.02515	-0.02475	-0.02634	-0.02575
a_5	-0.00874	-0.00685	-0.00659	-0.00783	-0.00722
a_6	-0.00081	0.00062	0.00081	-0.00024	0.00038
a_7	-0.00149	-0.00034	-0.00020	-0.00114	-0.00051
a_8	-0.00253	-0.00156	-0.00145	-0.00231	-0.00168
a_9	-0.00119	-0.00033	-0.00025	-0.00106	-0.00042
a_{10}	0.00004	0.00080	0.00087	0.00009	0.00073
b_1	0.69274	0.78166	0.78887	0.59177	0.78603
b_2	-0.02558	0.02081	0.01770	-0.06287	0.01763
b_3	-0.14357	-0.11254	-0.11512	-0.16728	-0.11505
b_4	-0.04623	-0.02303	-0.02509	-0.06383	-0.02502
b_5	-0.03176	-0.01321	-0.01490	-0.04575	-0.01483
b_6	-0.00825	0.00720	0.00578	-0.01987	0.00584
b_7	-0.00756	0.00567	0.00446	-0.01749	0.00451
b_8	-0.00808	0.00351	0.00243	-0.01675	0.00248
b_9	-0.00398	0.00632	0.00535	-0.01169	0.00539
b_{10}	-0.00410	0.00513	0.00425	-0.01107	0.00429

The worst case mean square errors of the reconstructed PPG signal of Fig. 6.1(a) for various values of k_{max} are tabulated in Table 6.2. Variation of the *NRMSE* as a percentage error in the reconstructed PPG as a function of k_{max} is also shown in Fig. 6.2. From Table 6.2 and Fig. 6.2, we can conclude that it is more than sufficient to compute and store only the first seven significant Fourier coefficients of each cycle to retain the morphological features of the given PPG signal with an accuracy of 0.5 %. Fig. 6.1(b) shows the reconstructed PPG signal using the first seven significant Fourier series coefficients. Once the time period and coefficients of a cycle are obtained, we can discard the original data set, as we can reconstruct the PPG from the time periods and the corresponding coefficients. The PPG in Fig. 6.1(a) was obtained with a sampling frequency of 200 Hz. Since after performing the proposed analysis, we need to keep only 16 values (T_m, $_m a_0$, $_m a_k \mid_{k=1,2,\ldots,7}$ and $_m b_k \mid_{k=1,2,\ldots,7}$) per cycle, a data compression factor of 12.5 is automatically achieved.

6.3 Cycle-by-cycle FSA for Motion Artifact Reduction

When a patient moves, the resultant corrupted PPG, say $f_c(t)$ will contain additional components due to motion artifact as $f_c(t) = f(t) + f_a(t)$, where $f_c(t)$ is the corrupted PPG, $f(t)$ is the original artifact-free PPG and $f_a(t)$ is the motion artifact. Now if we compute the Fourier series coefficients of the mth cycle of $f_c(t)$, we get:

$$_m a_{0c} = \frac{1}{T_m} \int_0^{T_m} f_c(t)\, dt = \frac{1}{T_m} \int_0^{T_m} [f(t) + f_a(t)]\, dt = a_0 + \frac{1}{T_m} \int_0^{T_m} f_a(t)\, dt \qquad (6.7)$$

$$_m a_{kc} = \frac{2}{T_m} \int_0^{T_m} f_c(t) \cos(k\,\omega_m t)\, dt = {}_m a_k + \frac{2}{T_m} \int_0^{T_m} f_a(t) \cos(k\,\omega_m t)\, dt \qquad (6.8)$$

Table 6.2 Error in reconstruction of PPG as a function of chosen number of Fourier coefficients for reconstruction. (It is easily seen that the optimum number of coefficients is 7)

No. of Coefficients	*NRMSE*	
	%	dB
1	13.19	-17.60
2	4.00	-27.96
3	2.02	-33.89
4	0.60	-44.42
5	0.59	-44.59
6	0.57	-44.82
7	0.51	-45.85
8	0.51	-45.90
9	0.50	-45.96
10	0.49	-46.28

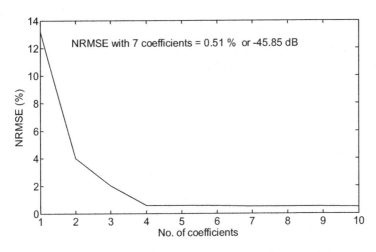

Fig. 6.2 *NMRSE* (%) as a function of number of Fourier coefficients

$$_mb_{kc} = \frac{2}{T_m}\int_0^{T_m} f_c(t)\sin(k\omega_m t)\,dt = {}_mb_k + \frac{2}{T_m}\int_0^{T_m} f_a(t)\sin(k\omega_m t)\,dt. \quad (6.9)$$

Since the cardiac synchronous PPG pulsation has very little correlation with the motion artifact signal, they can be assumed to be independent of each other [80]. Consequently the terms $\frac{2}{T_m}\int_0^{T_m} f_a(t)\cos(k\omega_m t)\,dt$ in equation (6.8) and $\frac{2}{T_m}\int_0^{T_m} f_a(t)\sin(k\omega_m t)\,dt$ in equation (6.9) would individually evaluate to zero. Hence under this condition we will have ${}_ma_{kc} = {}_ma_k$ and ${}_mb_{kc} = {}_mb_k$. Furthermore, the motion artifacts in general are random in nature and can be assumed to have zero mean value. From equation (6.7), it is easily seen that if the artifact $f_a(t)$ has no DC value then ${}_ma_{0c} = {}_ma_0$ itself. Thus, if we reconstruct the PPG from the computed Fourier series coefficients of the corrupted PPG, we obtain the original (artifact-free) PPG.

In order to demonstrate the efficacy of the proposed method, a known motion artifact is intentionally added to the clean PPG of Fig. 6.1(a). The resulting corrupted PPG is indicated in Fig. 6.3(a). The proposed cycle-by-cycle FSA is then performed on the PPG of Fig. 6.3(a) restricting the number of Fourier coefficients to 7 and the reconstructed PPG is given in Fig. 6.3(b). Fig. 6.3(b) clearly indicates that the method extracts clean artifact-free PPG from a PPG corrupted with motion artifacts. A visual inspection of Fig. 6.3(b) does not indicate the presence of any artifact. However, the residual of the a priori known added motion artifact embedded in the PPG of Fig. 6.3(b) was extracted utilizing the synchronous detection technique. The ratio of the residual motion artifact in the PPG of Fig. 6.3(b) to that of the original level of artifact in the PPG of Fig. 6.3(a)

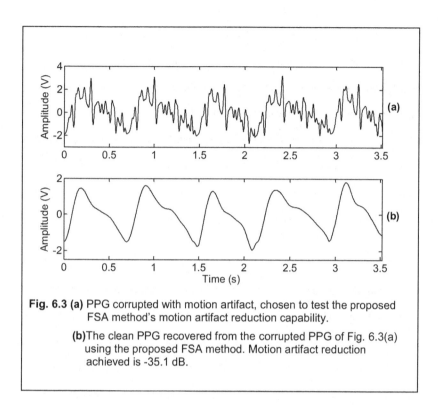

Fig. 6.3 (a) PPG corrupted with motion artifact, chosen to test the proposed FSA method's motion artifact reduction capability.

(b) The clean PPG recovered from the corrupted PPG of Fig. 6.3(a) using the proposed FSA method. Motion artifact reduction achieved is -35.1 dB.

provides the reduction factor achieved with the proposed method. The reduction in the motion artifact is observed to be -35.1 dB.

6.4 Experimental Results

To verify the practicality of the proposed FSA technique for motion artifact reduction and data compression, necessary hardware for sensing and software for acquiring and processing the red and IR PPG signals are developed. The analog front end developed for driving the sensor head, portrayed in chapter 2 is employed for this task as well. As in the earlier case, here too, the data acquisition and processing were accomplished under the LabVIEW environment. Forty sets of PPG data were obtained from volunteers with intentionally induced frequently encountered artifacts, namely, (i) vertical motion, (ii) horizontal motion, (iii) bending of the finger and (iv) applying pressure on the sensor (pressing the finger) following the protocol as described in the previous chapter. The acquired data from the volunteers were first filtered using the Savitzky-Golay (SG) smoothing filter to remove high frequency noise as SG filters are optimal and provide minimum least-square error in fitting a polynomial to noisy data. The individual cycles were identified and the proposed algorithm operated on each identified cycle. Fourier coefficients $_m a_{0_R}$, $_m a_{k_R}$ and $_m b_{k_R} \big|_{k_R=1,2,3..k_{R_{max}}}^{m=1,2,3...}$ for the red PPG and coefficients $_m a_{0_{IR}}$, $_m a_{k_{IR}}$ and $_m b_{k_{IR}} \big|_{k_{IR}=1,2,3..k_{IR_{max}}}^{m=1,2,3...}$ for the IR PPG ascertained there from. Each cycle of the PPG is then reconstructed from the first seven coefficients employing equation (6.5). The level of oxygen saturation (SpO$_2$) was also computed using the measurements from the raw PPG signals and the reconstructed PPG signals utilizing equation (4.5). The proposed algorithm performed well in each case providing a reduction in motion artifact between 35 dB and 40 dB. The quantification of reduction is obtained by taking as reference the two one minute artifact-free PPG portions, preceding and succeeding the artifact ridden two-minute portion, of every recording.

One of the difficulties with objectively evaluating any algorithm for artifact reduction is that there is no benchmark of PPG signals available to test and compare the relative performances of one method over another. Most of the proposed algorithms are proprietary and their performances heavily depend on the settings and processing used by the manufacturers [72]. Therefore, in order to ascertain the effectiveness of motion artifact reduction, the performance of the proposed method is compared with the traditional moving average (MA) method of order 20 in each case of intentionally created motion artifact. Fig. 6.4, portrays

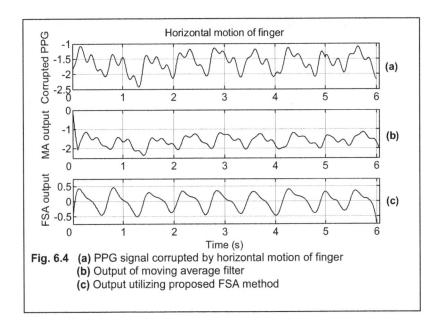

Fig. 6.4 (a) PPG signal corrupted by horizontal motion of finger
(b) Output of moving average filter
(c) Output utilizing proposed FSA method

the results for the motion artifact created by the horizontal motion of the finger. Fig. 6.5 shows the artifact created by simply pressing the finger and the sensor

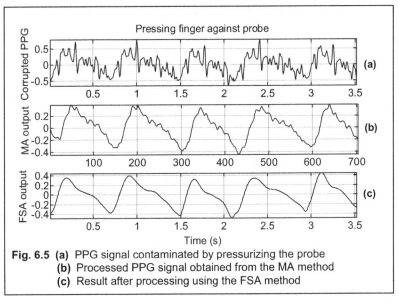

Fig. 6.5 (a) PPG signal contaminated by pressurizing the probe
(b) Processed PPG signal obtained from the MA method
(c) Result after processing using the FSA method

head connected and its efficient removal by the proposed FSA method. Fig. 6.6 illustrates the efficacy of the proposed method for artifacts due to vertical motion of the finger and Fig. 6.7 demonstrates the proposed method's ability to reduce the artifacts created by the bending action of the finger. A visual inspection of all the figures shows that, in all the cases, the artifacts are significantly reduced by the proposed methodology compared to the MA filter. Moreover, the SpO_2 readings obtained after artifact reduction by the proposed FSA method during the period of motion closely matches with the ones just prior to the on-set of motion, again establishing the practical applicability of the proposed method.

As the estimation of oxygen saturation as per equation (4.5) depends on the peak-to-peak amplitudes the PPG signal, it is important that the reduction method proposed here preserves this characteristic in the recovered PPG. Hence the peak-to-peak values of the artifact suffered raw PPG and the PPG recovered after processing using the proposed FSA technique are analyzed and the results of the statistical analysis are tabulated in Table 6.3, in terms of mean and standard deviation (SD). Here too the measurements obtained on the first one minute of artifact-free PPG are taken as the base line reference for further computations and comparison. It is seen from Table 6.3 that the FSA method performs extremely well in restoring the peak-to-peak amplitude of the corrupted PPG cycles and provides better performance than the MA filter. Furthermore, it can be observed that the proposed method restores the important morphological features of the PPG. In each case, SpO_2 values were also computed.

Since most commercial pulse oximeters suppress the reading during motion episodes and are incapable of operating correctly during artifacts, the comparison is made between the SpO_2 value computed from the artifact-free one minute precursor and tail end portions of the PPG and the SpO_2 computed utilizing the

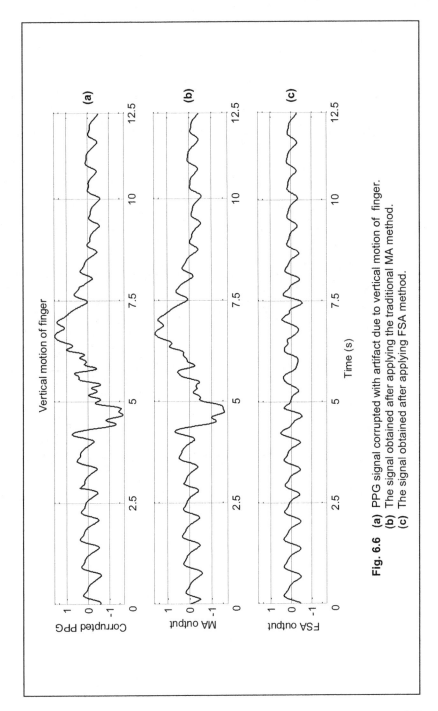

Fig. 6.6 (a) PPG signal corrupted with artifact due to vertical motion of finger.
(b) The signal obtained after applying the traditional MA method.
(c) The signal obtained after applying FSA method.

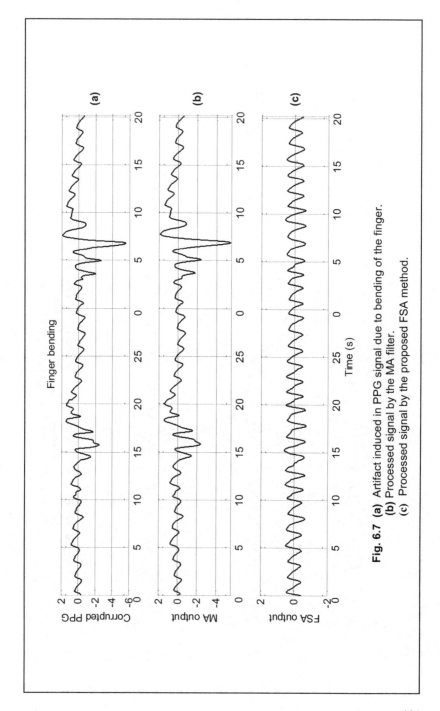

Fig. 6.7 (a) Artifact induced in PPG signal due to bending of the finger. **(b)** Processed signal by the MA filter. **(c)** Processed signal by the proposed FSA method.

Table 6.3 Effectiveness of the FSA method in restoring the peak-to-peak values of a PPG

PPG afflicted with different artifacts	Vertical motion	Horizontal motion	Bending motion	Excess pressure
A clean section of PPG	0.79 ± 0.09	0.78 ± 0.09	0.88 ± 0.11	0.75 ± 0.02
Artifact corrupted PPG	0.88 ± 0.32	0.87 ± 0.12	2.24 ± 2.04	1.23 ± 0.14
PPG recovered by MA filter	0.66 ± 0.30	0.80 ± 0.10	1.32 ± 1.31	0.69 ± 0.08
PPG recovered by FSA	0.74 ± 0.16	0.78 ± 0.09	0.89 ± 0.13	0.78 ± 0.04

The data indicates the mean ± SD of the peak-to-peak values (V) of the raw PPG, processed PPG with the MA and the proposed FSA methods.

artifact ridden two minute portion of the PPG. While the moving average method never gave reliable results, the SpO_2 values computed after applying artifact reduction with the proposed FSA method are always within 3% of the expected values (values obtained during the artifact-free recording). To illustrate this aspect of the analysis, a sample of red and IR PPG signals corrupted with motion artifact (with a 10 s artifact free signal preceding the corrupted portion) are shown in Fig. 6.8(a) and Fig. 6.8(c) respectively. The reconstructed red and IR PPG signals employing the proposed cycle-by-cycle FSA method are indicated in Fig. 6.8(b) and Fig. 6.8(d) respectively. SpO_2 estimated employing equation (4.5) utilizing the raw PPG signals and the processed PPG signals are given in Fig. 6.8(e). It is seen from the first 10 s of the PPG that the estimated SpO_2, after artifact reduction employing the proposed technique matches closely with the ones obtained from the raw signal when the PPG signals are not corrupted indicating that the proposed method does not introduce errors due to the

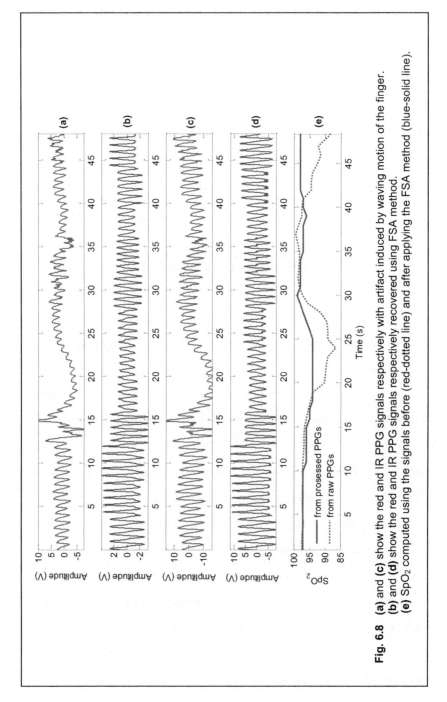

Fig. 6.8 (a) and (c) show the red and IR PPG signals respectively with artifact induced by waving motion of the finger. (b) and (d) show the red and IR PPG signals respectively recovered using FSA method. (e) SpO_2 computed using the signals before (red-dotted line) and after applying the FSA method (blue-solid line).

additional processing of the PPG signals. The efficacy of artifact reduction capability of the proposed method is seen in the region of 10 s to 45 s in Fig. 6.8. While the estimation of SpO_2 from the artifact corrupted PPG signals is way off with errors more than 25 % from the expected values for the moving average method, the SpO_2 computed using the PPG signals processed by the proposed cycle-by-cycle FSA technique is much more stable and is always within 3 % of the values indicated before the onset of artifacts. The estimated SpO_2 before and after processing the PPGs corrupted by the artifacts, are presented in the Annexure 4.

7. Summary and Conclusions

7.1 Summary of the Work Presented in this Book

Over the past decade, the oxygen content in arterial blood (SaO_2) has emerged as the fifth vital sign indicating the status of health of a patient, in addition to the traditional vital signs, namely, body temperature, blood pressure, pulse rate and respiratory rate. Pulse oximeters, utilizing a couple of photoplethysmographs, estimate both pulse rate and SaO_2 (indicated as SpO_2 in pulse oximeter vocabulary). However, pulse oximeters in vogue have several limitations that may lead to inaccurate readings. Hence pulse oximeters are extensively calibrated to remove the effect of patient and sensor dependent parameters on the computation of SpO_2. The work presented in this book addresses this issue in the main and presents three novel methods for computation of SpO_2 estimation, which are relatively insensitive to patient dependent parameters such as, skin pigmentation, volume of the tissue and also to the source and the detector sensitivities. All the three methods proposed here do not rely on calibration curves, obtained through empirical curve fitting, as is normally done in present day pulse oximeters.

Movement of a patient connected to the pulse oximeter causes large fluctuations in the PPG signals, resulting in erroneous SpO_2 computations. Addressing this issue, a couple of signal processing methods suited for the reduction of motion artifacts from PPG signals are also presented, with a view to improve the reliability of SpO_2 estimations.

All the research works reported in this book are supported by theoretical derivations and validation through experimentation on suitable prototypes built and tested.

7.2 Conclusions

(i) The Photoplethysmograph designed and developed employing negative feedback compensation scheme normalizes the DC component of a PPG signal. This scheme not only compensates for the color of the skin but also compensates for the intervening volume of tissue coming in the path of light and thus makes the PPG signals patient and sensor independent. Though this novel model based method provides clinically acceptable SpO_2 estimations within ± 2 % of the ones indicated by a commercial pulse oximeter, the feedback makes oximeter sensitive to motion artifacts.

(ii) The slope based method, derived from a model for light propagation through the test specimen (finger or earlobe as the case may be), though provides an excellent method of computation of SpO_2, requires complex processing steps involving extensive computations.

(iii) The peak value method derives its strength from the slope based method, but possesses reduced number of processing steps leading to reduced level of computation. Hence is the best of the three methods proposed here and is better suited for "on-line" measurement of SpO_2. Annexure 3 provides a comparison of all the three methods outlined in the book.

(iv) The signal processing technique based on SVD that can effectively reduce the artifacts in PPG signal caused by movement of a patient yields a significant improvement in motion artifact rejection. However, the method requires computation of SVR spectra and hence is computationally intensive. The method is best suited for "off-line" processing.

(v) A computationally efficient method for motion artifact reduction from a PPG signal, by applying the well known Fourier series analysis (FSA) on a cycle-by-cycle basis presented here gave a motion artifact reduction of -35dB to -40dB at the same time provided a data compression factor of 12.5. This method is applicable for "on-line" processing of PPG signals. Annexure 4 illustrates that the proposed method provides stable SpO_2 readings, even during periods wherein artifacts are present in the PPG signals.

Portions of the work presented in the book have been published in the form of academic publications and some are filed for patenting in the Indian patents office. The details are given below.

Patents (Filed)

1. Non-invasive method and device for the measurement of oxygen saturation in arterial blood - I (No. 1189/CHE/2007).
2. Non-invasive method and device for the measurement of oxygen saturation in arterial blood - II (No. 1191/CHE/2007).

International Conferences

[1] **K. A. Reddy**, J. R. Bai, B. George, N. M. Mohan and V. J. Kumar, "Virtual instrument for the measurement of haemo-dynamic parameters using photoplethysmograph," in Proc. 23rd Int. Conf. IEEE, IMTC-2006, Sorrento, Italy,24-27 April, 2006, pp. 1167-1171. (DOI: 10.1109/IMTC.2006.328443)

[2] **K. A. Reddy** and V. J. Kumar, "Motion artifact reduction in photoplethysmographic signals using singular value decomposition," in Proc. 24th Int. Conf. IEEE, IMTC-2007, Warsaw, Poland, 1–3 May, 2007, pp. 1-5. (DOI: 10.1109/IMTC.2007.379467)

[3] **K. A. Reddy**, B. George and V.J. Kumar, "Motion artifact reduction and data compression of photoplethysmographic signals utilizing cycle-by-cycle fourier series analysis", in Proc. 25rd Int. Conf. IEEE, I2MTC-2008, Victoria, Canada,12-15 May, 2008, pp. 176-179. (DOI: 10.1109/IMTC.2008.4547026)

[4] **K. A. Reddy**, B. George , N.M. Mohan and V.J.Kumar, "A Novel method of measurement of oxygen saturation in arterial blood", in Proc. 25rd Int. Conf. IEEE, I2MTC-2008, Victoria, Canada,12-15 May, 2008, pp. 1627-1630. (DOI: 10.1109/IMTC.2008.4547304)

Journal

[1] **K. A. Reddy**, B. George and V. J. Kumar, "Use of Fourier analysis for motion artifact reduction and data compression of photoplethysmographic signals", *IEEE Trans. Instrum. Meas.,* vol. 58, no.5, pp.1706-1711, May 2009. (DOI: 10.1109/TIM.2008.2009136)

[2] **K. A. Reddy**, B. George, N.M. Mohan and V. J. Kumar, "A Novel calibration–free method of measurement of oxygen saturation in arterial blood", IEEE Trans. Instrum. Meas., vol. 58, no.5, pp.1699-1705, May 2009. (DOI: 10.1109/TIM.2008.2012934)

7.3 Scope for Future Work

The model for the propagation of light given in this book assumes the scattered light as part of the attenuation. Though the assumption works well, it would be advantageous and may lead to more insights and simpler method of computation of SpO_2, if a model incorporating scattering is developed and studied.

In this study, to validate the proposed SpO_2 estimation methods, data from healthy volunteers are used for trials. And hence the range for testing is limited to 96% to 99%. The validity of the proposed methods outside this range of oxygen saturation can be further studied.

References

[1] E. N. Bruce, *Biomedical Signal Processing and Signal Modeling*, John Wiley & Sons, NY, 2001.

[2] R. M. Rangayyan, *Biomedical Signal Analysis: A Case Study Approach*, IEEE Press Series on Biomedical Engineering, John Wiley & Sons, Singapore, 2002.

[3] L. S. Sornmo and P. Laguna, *Bioelectrical Signal Processing in Cardiac and Neurological Applications*, Elsevier Academic Press, USA, 2005.

[4] W. F. Ganong, *Review of Medical Physiology*, 16th edition, Appleton & Lange, Norwalk, CT, 1993.

[5] D. L. Kasper, *Harrison's Principles of Internal Medicine*, 16th edition, McGraw-Hill, NY, 2005.

[6] John K-J Li, *Dynamics of Vascular System*, Series on Bioengineering & Biomedical Engineering, vol. 1, World Scientific Publishing Co. Pte. Ltd, Singapore, 2004.

[7] R. D. Miller, *Miller's Anesthesia*, 6th edition, Elsevier Churchill Livingstone, Philadelphia, 2005.

[8] J. W. Severinghaus and P. B. Astrup, "History of blood gas analysis-VI. Oximetry," *J. Clin. Monit.*, vol. 2, no. 4, pp. 270-88, 1986.

[9] J. G. Webster, *Design of Pulse Oximeters*, Taylor & Francis Group, NY, 1997.

[10] J. A. Dorsch and S. E. Dorsch, *Understanding Anaesthesia Equipment*, Williams & Wilkins, Baltimore, 1999.

[11] G. D. Baura, *System Theory and Practical Applications of Biomedical Signals*, IEEE press series on biomedical engineering, John Wiley & Sons, NJ, 2002.

[12] T. Ahrens and K. Rutherford, *Essentials of Oxygenation*, Jones & Barlett, Boston, 1993.

[13] Y. Pole, "Evolution of the pulse oximeter," *International Congress Series*, vol. 1242, pp. 137-142, 2002.

[14] C. Secker and P. Spiers, "Accuracy of pulse oximetry in patients with low systematic vascular resistance," *Anaesthesia*, vol. 52, no. 2, pp. 127-130, 1997.

[15] A. J. Williams, "ABC of oxygen: Assessing and interpreting arterial blood gases and acid-base balance," *BMJ*, vol. 317, pp. 1213-1216, 1998.

[16] A. B. Hertzman, "The blood supply of various skin areas as estimated by the photoelectric plethysmograph," *Am. J. Physiol.*, vol. 124, pp. 328-340, 1938.

[17] K. Nakajima, T. Tamura and H. miike, "Monitoring of heart and respiratory rates by Photoplethysmography using a digital filtering technique," *Med. Eng. Phys.*, vol. 18, no. 5, pp. 365-372, 1996.

[18] V. Blazek and U. Schultz-Ehrenburg, *Quantitative Photoplethysmography: Basic facts and examination tests for evaluating peripheral vascular functions,* VDI Verlog, 20(192), 1996.

[19] S. Sarin, D. A. Shields, J. H. Scurr and P. D. C. Smith, "Photoplethysmography: a valuable noninvasive tool in the assessment of venous dysfunction ?," *J. Vasc. Surg.*, vol. 16, no. 2, pp. 154-162, 1992.

[20] G. A. Millikan, "The oximeter, an instrument for measuring continuously the arterial saturation of arterial blood in man," *Rev. Sci. Instrum.*, vol. 13, pp. 434 – 444, 1942.

[21] E.A.G. Goldie, "Device for continuous indication of oxygen saturation of circulating blood in man," *J. Sci. Instrum.*, vol. 19, pp. 23– 5, 1942.

[22] E. H. Wood and J. E. Geraci, "Photoelectric determination of arterial oxygen saturation in man," *J. Lab. Clin. Med.*, vol. 34, pp. 387-401, 1949.

[23] R. Brinkman and W. G. Zijlstra, "Determination and continuous registration of the percentage oxygen saturation in clinical conditions," *Arch. Chir. Neerl.*, vol. 1, pp. 177-183, 1949.

[24] P. Sekelj, A. L. Johnson, H. E. Hoff and P. M. Scherch, "A photoelectric method for the determination of arterial oxygen saturation in man," *Amer. Heart. J.*, vol. 42, pp. 826-848, 1951.

[25] M. L. Polanyi and R. M. Hehir, "New reflection oximeter," *Rev. Sci. Instru.*, vol. 31, no. 4, pp. 401-403, 1960.

[26] A. Cohen and N. A. Wardsworth, "Light emitting diode skin reflectance oximeter," *Med. Biol. Eng.*, vol. 10, pp. 385-391, 1972.

[27] T. Aoyagi, M. Kishi, K. Yamaguchi, and S. Watanabe, "Improvement of an ear-piece oximeter," in *Proc. Abstracts of 13th Annu. Japanese Soc. Med. Electro. Biologic Eng.*, JSMEBE, Osaka, Japan, pp. 90-91, 1974.

[28] E. B. Merrick and T. J. Hayes, "Continuous, non-invasive measurements of arterial blood oxygen levels," *Hewlett-packard J.*, vol. 28, no. 2, pp. 2-9, 1976.

[29] S. Takatani and J. Ling, "Optical oximetry sensors for whole blood and tissue," *IEEE Eng. Med. Biol. Mag.*, pp. 347-357, June/July 1994.

[30] R. J. Falconer and B. J. Robinson, "Comparison of pulse oximeters: accuracy at low arterial pressure in volunteers," *BJA*, vol. 65, no. 4, pp. 552-557, 1990.

[31] F. W. Cheney, "The ASA closed claims study after the pulse oximeter," *ASA Newsletter*, vol. 54, 1990.

[32] B. L. Horecker, "The absorption spectra of hemoglobin and its derivatives in the visible and near infra-red regions," *J. Biol. Chem.*, vol. 148, no. 1, pp. 173-183, 1943.

[33] W. G. Zijlstra, A. Buursma and W. P. Meeuwsen-van der Roest, "Absorption spectra of human fetal and adult oxyhemoglobin, de-

oxyhemoglobin, carboxyhemoglobin and methemoglobin," *Clin. Chem.,* vol. 37, no.9, pp. 1633-1638, 1991.

[34] J. G. Kim, M. Xia and H. Liu, "Extinction coefficients of hemoglobin for near-infrared spectroscopy of tissue", *IEEE Eng. Med. Biol. Mag.,* vol. 24, no. 2, pp. 118-121, March/April 2005.

[35] Y. Mendelson, "Pulse oximetry: theory and applications for noninvasive monitoring," *Clin. Chem.,* vol. 38, no. 9, pp. 1601-1607, 1992.

[36] J. E. Sinex, "Pulse oximetry: Principles and limitations," *Am. J. Emerg. Med.,* vol. 17, no. 1, pp. 59-68, 1999.

[37] M. W. Wukitsch, M. T. Petterson, D. R. Tobler and J. A. Pologe, "Pulse oximetry: Analysis of theory, technology, and practice," *J. Clin. Monit.,* vol. 4, no. 4, pp. 290-301, 1988.

[38] T. L. Rusch, R. Sankar and J. E. Scharf, "Signal processing methods for pulse oximetry," *Comput. Biol. Med.,* vol. 26, no. 2, pp. 143-159, 1996.

[39] M. Brown and J. S. Vender, "Noninvasive oxygen monitoring," *Crit. Care. Clin.,* vol. 4, no. 3, pp. 493-509, 1988.

[40] W. A. Bowes 3rd, B. C. Corke and J. Hulka, "Pulse oximetry: a review of the theory, accuracy, and clinical applications," *Obstet. Gynecol.,* vol. 74 (3 Pt 2), pp. 541-546, 1989.

[41] A. Huch, R. Huch, V. Konig, M. R. Neuman, D. Parker, J. Yount and D. Lubbers, "Limitations of pulse oximetry (letter)," *Lancet.,* vol. 1, no. 8581, pp. 357-358, 1988.

[42] D. Amar, J. Neidzwski, A. Wald and A. D. Finck, "Fluorescent light interferes with pulse oximetry," *J. Clin. Monit.,* vol. 5, no. 2, pp. 135-136, 1989.

[43] C. J. Cote, E. A. Goldstein, W. H. Buschman and D. C. Hoaglin, "The effect of nail polish on pulse oximetry," *Anesth. Analg.,* vol. 67, no. 7, pp. 683-686, 1988.

[44] M. M. Chan, M. M. Chan and E. D. Chan, "What is the effect of finger nail polish on pulse oximetry?," *Chest*, vol. 123, no. 6, pp. 2163-2164, 2003.

[45] A. L. Ries, L. M. Prewitt and J. J. Johnson, "Skin color and ear oximetry," *Chest*, vol. 96, no. 2, pp. 287-290, 1989.

[46] J. A. Langton and C. D. Hanning, "Effect of motion artefact on pulse oximeters: evaluation of four instruments and finger probes," *Br. J. Anaesth.*, vol. 65, pp. 564-570, 1990.

[47] J. L. Plummer, A. Z. Zakaria, A. H. Ilsley, R. R. L. Fronsko and H. Owen, "Evaluation of the influence of movement on saturation readings from pulse oximeters," *Anaesthesia*, vol. 50, pp. 423-426, 1995.

[48] S. J. Barker, K. K. Tremper and J. Hyatt, "Effects of methemoglobinemia on pulse oximetry and mixed venous oximetry," *Anesthesiology*, vol. 70, no. 1, pp. 112-117, 1989.

[49] W. P. Bozeman, R. A. Myers and R. A. Barish, "Confirmation of pulse oximetry gap in carbonmonoxide poisoning," *Ann. Emerg. Med.*, vol. 30, pp. 608-611, 1997.

[50] A. Sidi, D. A. Paulus, Rush W, N. Gravenstein and R. F. Davis, "Methylene blue and indocyanine green artifactually lower pulse oximetry readings of oxygen saturation: Studies in dogs," *J. Clin. Monit.*, vol. 3, no. 4, pp. 249-256, 1987.

[51] M. S. Scheller, R. J. Unger and M. J. Kelner, "Effects of intravenously administered dyes on pulse oximetry readings," *Anesthesiology*, vol. 65, no. 5, pp. 550-552, 1986.

[52] L.M. Schnapp and N.H. Cohen, "Pulse oximetry: uses and abuses," *Chest*, vol. 98, no. 5, pp. 1244-1250, 1990.

[53] N. S. Trivedi, A. F. Ghouri, N. K. Shah, E. Lai and S. J. Barker, " Effects of motion, ambient light and hypoperfusion on pulse oximeter function," *J. Clin. Anesth.*, vol. 9, no. 3, pp. 179–183, 1997.

[54] R. R. Fluck Jr, C. Schroeder, G. Frani, B. Kropf and B. Engbretson, "Does ambient light affect the accuracy of pulse oximetry?," *Resp. care.*, vol. 48, no.7, pp. 677-680, 2003.

[55] P. F. White and W. A. Boyle, "Nail polish and oximetry," *Anesth. Analg.*, vol. 68, no. 4, pp. 546-547, 1989.

[56] C. D. Hanning and J. M. Alexander-Williams, "Pulse oximetry: a practical review," *BMJ*, vol. 311, pp. 367-370, 1995.

[57] A. Jubran, "Pulse oximetry," *Crit. Care.*, vol. 3, no. 2, pp: R11-R17, 1999.

[58] J. W. Saylor, "Neonatal and pediatric pulse oximetry," *Respir. Care.*, vol. 48, no. 4, pp. 386-98, 2003.

[59] K. A. Reddy, J. R. Bai, B. George, N. M. Mohan and V. J. Kumar, "Virtual instrument for the measurement of haemo-dynamic parameters using photoplethysmograph," *Proc. 23rd Int. Conf. IEEE, IMTC-2006,* Sorrento, Italy, pp. 1167-1171, 24-27 April, 2006.

[60] Y. Mendelson and B. D. Ochs, "Noninvasive pulse oximetry utilizing skin reflective photoplethysmography," *IEEE Trans. Biomed. Eng.*, vol. 35, no. 10, pp. 798-805, 1988.

[61] A. A. R. Kamal, J. B. Harness, G. Irving and A. J. Mearns, "Skin photoplethysmography - a review," *Comput. Methods. Prog. Biomed.*, vol. 28, pp. 257-269, 1989.

[62] S. Sugino, N. Kanaya, M. Mizuuchi, M. Nakayama and A. Namiki, "Forehead is as sensitive as finger pulse oximetry during general anesthesia," *Can. J. Anesth.*, vol. 51, pp. 432–436, 2004.

[63] J. A. Dean, *Lange's hand book of chemistry*, 14th edition, McGraw-Hill, NY, 1992.

[64] Y. Mattley, G. Leprac, R. Potter and L. Garcia-Rubio, "Light absorption and scattering model for the quantitative interpretation of human blood spectral data", *Photochemistry and Photobiology*, vol. 71, no. 5,pp. 610-619, 2000.

[65] M. Meinke, G. Muller, J. Helfmann and M. Friebel, "Optical properties of platelets and blood plasma and their influence on the optical behavior of whole blood in the visible to near infrared wavelength range", *J. Biomed. Opt.* vol. 12, no. 1, 014024, 2007.

[66] D. Damianou and J. A. Crowe, "The wavelength dependence of pulse oximetry", *IEE Colloquium on Pulse oxmetry: A critical appraisal*, pp. 7/1-7/3, 29 May 1996.

[67] http://sine.ni.com/nips/cds/view/p/lang/en/nid/14040

[68] S. J. Barker and N. K. Shah, "The effects of motion on the performance of pulse oximeter in volunteers," *Anesthesiology*, vol. 85, no. 4, pp. 101-108, 1996.

[69] A. B. Barreto, L. M. Vicente and I. K. Persad, "Adaptive cancellation of motion artifact in photoplethysmographic blood volume pulse measurements for exercise evaluation," in *Proc. IEEE-EMBC and CMBEC*, vol. 2, pp. 983-984, 20-23 sept.,1995.

[70] A. R. Relente and L. G. Sison, "Characterization and adaptive filtering of motion artifacts in pulse oximetry using accelerometers", in *Proc. Conf. EMGS/BMES*, Houston, TX, USA, pp. 1769-1770, 23-26 Oct., 2002.

[71] K. W. Chan and Y. T. Zhang, "Adaptive reduction of motion artifact from photoplethysmographic recordings using a variable step-size LMS filter," in *Proc. IEEE Sensors*, vol. 2, pp. 1343- 1346, 2002.

[72] F. M. Coetzee and Z. Elghazzawi, "Noise-resistant pulse oximetry using a synthetic reference signal," *IEEE trans. Biomed. Engg.*, vol. 47, no. 8, pp. 1018-1026, 2000.

[73] J. M. Goldman, M. T. Petterson, R. J. Kopotic and S. J. Barker, "Masimo signal extraction pulse oximetry," *J. Clin. Monit. Comput.*, vol. 16, no. 7, pp. 475-483, 2000.

[74] J. Lee, W. Jung, I. Kang, Y. Kim and G. Lee, "Design of filter to reject motion artifact of pulse oximetry," *Computer Standards and Interfaces*, vol. 26, No. 3, pp. 241-249, 2004.

[75] C. M. Lee and Y. T. Zhang, "Reduction of motion artifacts from photoplethysmographic recordings using a wavelet denoising approach," in *Proc. IEEE EMBS Asian-Pacific Conf. on Biomed. Engg.*, pp. 194-195, 20-22 Oct., 2003.

[76] Y. S. Yan, C. C. Poon and Y. T. Zhang, "Reduction of motion artifact in pulse oximetry by smoothed pseudo wigner-ville distribution," *J Neuroengineering. Rehabil.*, 2005, 2:3, available from http://www.jneuroengrehab.com/content/2/1/3.

[77] M. J. Hayes and P. R. Smith, "Artifact reduction in photoplethysmography," *Appl. Opt.*, vol. 37, no. 31, pp. 7437–7446, 1998.

[78] M. J. Hayes and P. R. Smith, "A new method for pulse oximetry possessing inherent insensitivity to artifact," *IEEE Trans. Biomed. Engg.*, vol. 48, No. 4, pp. 452-461, 2001.

[79] P. F. Stetson, "Independent component analysis of pulse oximetry signals," in *Proc. 26th Annual Intern. Conf. IEEE EMBS*, San Francisco, CA, USA, pp. 231-234, 1-5 Sept., 2004.

[80] B. S. Kim and S. K. Yoo, "Motion artifact reduction in photoplethysmography using independent component analysis," *IEEE Trans. Biomed. Engg.*, vol. 53, No. 3, pp. 566- 568, 2006.

[81] J. Yao and S. Warren, "A short study to assess the potential of independent component analysis for motion artifact separation in wearable pulse oximeter signals," in *Proc. 27th Annual Conf. IEEE Engg. Medicine and Biology,* Shanghai, China, pp. 3585-3588, 1-4 Sept., 2005.

[82] J. Y. A. Foo, "Comparison of wavelet transformation and adaptive filtering in restoring artifact-induced time-related measurement," *Biomedical signal processing and control*, vol. 1, Issue. 1, pp. 93-98, 2006.

[83] K. A. Reddy and V. J. Kumar, "Motion artifact reduction in photoplethysmographic signals using singular value decomposition," in *Proc. 24th Int. Conf. IEEE, IMTC-2007,* Warsaw, Poland, pp. 1-4, 1–3 May, 2007.

[84] V. C. Klema and A. J. Laub, "The singular value decomposition: Its computation and some applications," *IEEE Trans. Autom. Control.*, vol. AC-25, no. 2, pp. 164-176, 1980.

[85] P. P. Kanjilal and S. Palit, "On multiple pattern extraction using singular value decomposition," *IEEE Trans. Signal. Process.*, vol. 43, no. 6, pp. 1536-1540, 1995.

[86] R. N. Bracewell, *The fourier transform and its applications*, 3rd edition, McGraw-Hill higher education, NY, 2000.

Annexure 1

As discussed in chapter 2, the red and IR PPG signals are frequency modulated and interfaced to a PC utilizing the audio channel. Block diagram of the vi used to acquire the FM modulated PPG data from sound card is shown in Fig. A1.1. Subsequently, block diagram of the vi shown in Fig. A1.2 demodulates the acquired signals, computes and displays the Pulse rate and SpO_2 readings.

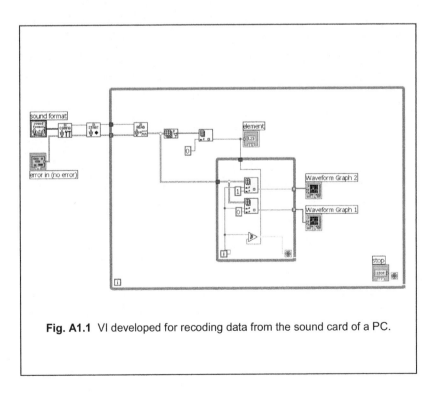

Fig. A1.1 VI developed for recoding data from the sound card of a PC.

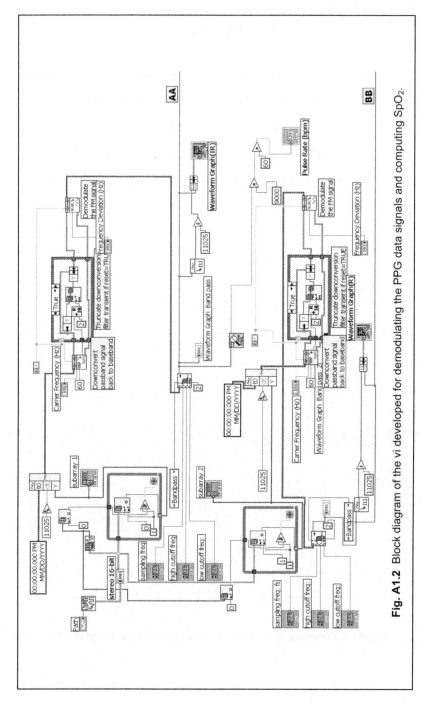

Fig. A1.2 Block diagram of the vi developed for demodulating the PPG data signals and computing SpO$_2$.

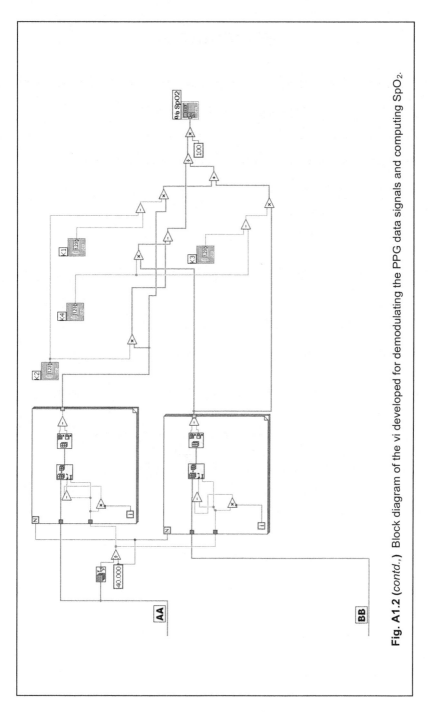

Fig. A1.2 (*contd.,*) Block diagram of the vi developed for demodulating the PPG data signals and computing SpO$_2$.

Annexure 2

As detailed in chapter 3, the red and IR PPG signals are sampled and acquired using a 16-bit data acquisition card NI DAQPad-6015 manufactured by National Instruments. Block diagram of the vi developed to acquire both red and IR PPG signals along with the representative DC signals is shown in Fig. A2.1. Block diagram of the vi developed to process the acquired signals and compute SpO_2 is portrayed in Fig. A2.2.

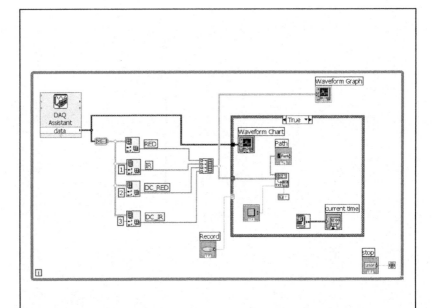

Fig. A2.1 VI developed to acquire the PPG data using National Instruments NI DAQPad-6015.

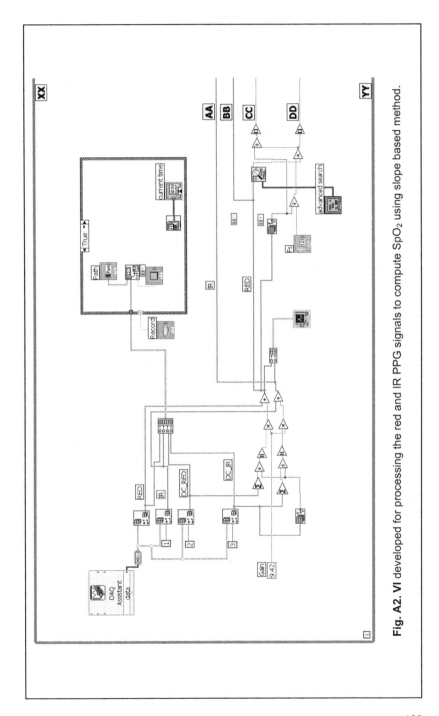

Fig. A2. VI developed for processing the red and IR PPG signals to compute SpO$_2$ using slope based method.

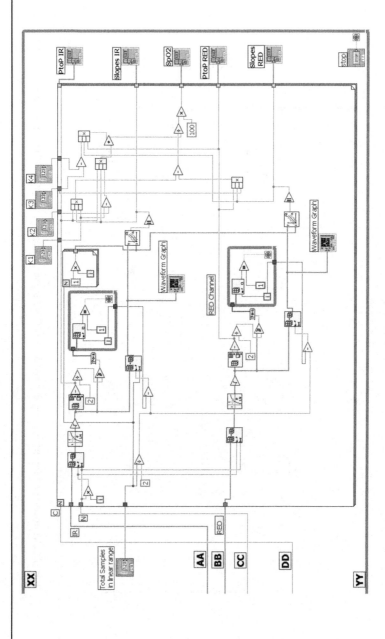

Fig. A2.2 (*contd.,*) VI developed for processing the PPG signals to compute SpO$_2$ using slope based method.

Annexure 3

The proposed methods of SpO$_2$ estimation, as detailed in chapter 2, 3, and 4 namely model based method, slope based method and peak value based method respectively, are applied on red and IR PPG signals. Test results on two subjects are presented in the form of SpO$_2$ plots as shown in Fig. A3.1.

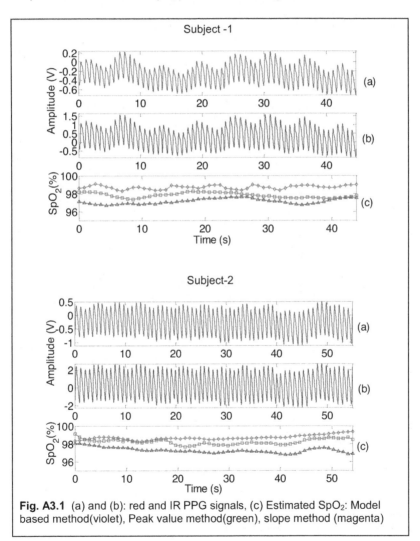

Fig. A3.1 (a) and (b): red and IR PPG signals, (c) Estimated SpO$_2$: Model based method(violet), Peak value method(green), slope method (magenta)

Annexure 4

As detailed in chapter 6, the proposed FSA method is applied on frequently encountered artifacts. The estimated SpO_2 before and after processing, utilizing equation (4.5), are presented in the Fig.A4.1. As seen from the Fig. A4.1, the proposed method performs well in recovering the corrupted PPGs there by enabling stable SpO_2 estimations during artifact periods.

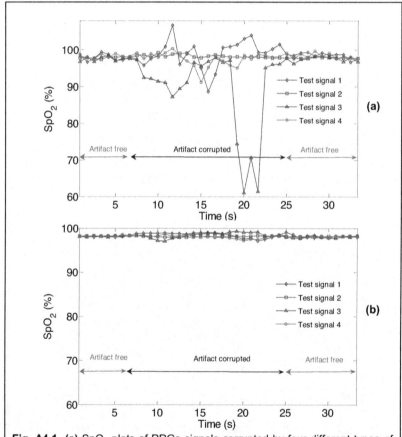

Fig. A4.1 (a) SpO_2 plots of PPGs signals corrupted by four different types of artifacts. The PPGs are free from artifacts during initial 7 s and last 10 s duration .The middle 7-25 s portions corresponds to artifact effected PPGs. **(b)** SpO_2 plots estimated after applying the FSA method on the PPGs.

I want morebooks!

Buy your books fast and straightforward online - at one of the world's fastest growing online book stores! Environmentally sound due to Print-on-Demand technologies.

Buy your books online at
www.get-morebooks.com

Kaufen Sie Ihre Bücher schnell und unkompliziert online – auf einer der am schnellsten wachsenden Buchhandelsplattformen weltweit!
Dank Print-On-Demand umwelt- und ressourcenschonend produziert.

Bücher schneller online kaufen
www.morebooks.de

OmniScriptum Marketing DEU GmbH
Heinrich-Böcking-Str. 6-8
D - 66121 Saarbrücken
Telefax: +49 681 93 81 567-9

info@omniscriptum.com
www.omniscriptum.com

Made in the USA
Middletown, DE
20 March 2020